Access All Areas

Access All Areas

A real world guide to gigging and touring

Trev Wilkins

Amsterdam • Boston • Heidelberg • London • New York • Oxford
Paris • San Diego • San Francisco • Singapore • Sydney • Tokyo

ELSEVIER

Focal Press is an imprint of Elsevier

Focal Press is an imprint of Elsevier
Linacre House, Jordan Hill, Oxford OX2 8DP, UK
30 Corporate Drive, Suite 400, Burlington, MA 01803, USA

First edition 2007

Notice
No responsibility is assumed by the publisher for any injury and/or damage to persons
or property as a matter of products liability, negligence or otherwise, or from any use
or operation of any methods, products, instructions or ideas contained in the material
herein. Because of rapid advances in the medical sciences, in particular, independent
verification of diagnoses and drug dosages should be made

British Library Cataloguing in Publication Data
A catalogue record for this book is available from the British Library

Library of Congress Cataloging-in-Publication Data
A catalog record for this book is available from the Library of Congress

ISBN–13: 978-0-24-052044-5
ISBN–10: 0-24-052044-0

For information on all Focal Press publications
visit our web site at books.elsevier.com

Printed and bound in Great Britain

07 08 09 10   10 9 8 7 6 5 4 3 2 1

Contents

Thanks and dedication

Special thanks go to the following people for their help in making this book happen . . .

All the good folks at Midas especially Richard Ferriday for help and support

Neutrik UK for technical information and samples

Shure microphones for technical information and diagrams

The technical team at Yamaha for technical information and support

All at Cakewalk software for their continued support and use of their products for screenshots

Sight 'n' Sound for giving me my first break

Stagecraft systems and Excel loudspeakers for help with photo subjects

Alex 'Bomber' Harris, backline technician extraordinaire, for getting me back out there

The crew at Coast to Coast productions, especially Vern, Jason, Gerard and Teaboy

Merv and the crew at Eurohire

Focal Press, especially my editor Catharine Steers for her help, support and belief in this project

Live Sound International magazine for their great mag

I've taken hundreds of photos on many shows for this book and would like to thank everyone who has been patient with me when they think I've gone mad because I'm taking a picture of a plug or the back of a mixer. A lot of people have also provided help and inspiration in numerous ways including ... Albert Lee and Hogan's Heroes,

Uriah Heep and crew, Limehouse Lizzy, Alexander O'Neal and band, Peter Andre and his brother Chris, Tom Robinson, House of Alice, Ian Hassel, Adam Farnsworth, Graham Ewins, Ian Wallace, Jakko Jacszyk, Robert Fripp, Mel Collins, Smokie, Styx and their crew, Double Cross, Bill Wyman's Rhythm Kings and Nick Griffiths who was an inspiration to me and is missed by many.

I'm sure there are people that I have forgotten to include but they know who they are. Thanks to all of you.

Dedication

Doing this job and living this lifestyle is only possible with the support and love of many special people who I would like to thank from the bottom of my heart. My Mother and Father who have always been there for me through good times and bad. My Sister Leanne and Brother Jon who always encourage me and look after everyone when I'm away. My children Damien, Daniel, Christian and Jamie who have put up with my traveling even when times were hard, and my new Grandson Rowan who is just starting his own journey through life.

A big thank you to all the artists and crew that I've worked with, particularly my 'road family' Albert Lee, Gerry Hogan, Peter Baron, Brian Hodgson, Pete Wingfield and Sue Hargreaves who have provided me with some wonderful times that I'll cherish forever.

My eternal love and thanks to Shirley who makes this life possible and waits patiently at home for me without a grumble or complaint.

1

Getting started

Welcome to the show

First of all, a big 'thank you' for purchasing this book. I do hope that you find it useful and keep it as a reference for when new situations arise or if you move into new fields within the industry. I'll be revising and updating the contents from time to time and will be adding new subjects along the way. If you have any requests for additional information, please let me know via the publisher.

The idea for the book developed as I came across many situations where information, knowledge, and communication would have made life much easier for the people involved in putting a show together, whether they be musicians, engineers, or venue staff. I don't envisage everyone being able to do every job required. Having some idea of how a show is put together from various viewpoints enables a deeper understanding of the bigger picture and also an appreciation of the problems faced by others.

The initial idea was to provide pointers for artists to improve their shows by understanding the equipment and techniques used by venues and crew in order to put on live productions. This idea was then extended to include an in-depth look at the 'nuts and bolts' involved so that potential engineers and technicians would be able to obtain a good grounding in many areas of live performance. This would also be relevant to artists wishing to further their technical knowledge, either to enable them to put together their own touring support system or so that they are better able to deal with crew and venues in a professional manner. It also provides a system of easy

(a)

(b)

Whether you're here (a) or here (b) the show must go on

communication tools that should be followed by artists, crew, and venues alike. This provides everyone with the correct information needed to perform at an optimum level using widely accepted documentation; forewarned is forearmed as they say.

I've tried to keep it as broadly based as possible. A theater glossary is included to clarify much of the language and terms used that are peculiar to it. Even if you aren't working in theater, you may find the glossary useful as many of the terms are used generally.

Wild child

I first got into this business (like most people) when I had an interest from my school years. I did some acting and musical projects but I was always fascinated by the 'behind the scenes' goings-on and, although I had my heroes, I didn't particularly want to be a rock star. I found myself taking an interest in amplifiers and speakers, which led me into going to a friend's band rehearsals where I started to balance their guitar amps so that they were all audible and not overpowering the drums or vocals. All I usually did was turn them down but this provided a much better sound for everyone and I then became 'sound engineer' for gigs! The out-front sound was just a couple of speakers with an amplifier for the microphone. There wasn't really any mixing to do after I'd balanced them during soundcheck, so I started to look into other things to do during the show.

I then went on the slippery slope of adding 'extras' into the show. This started with a few lights and soon developed to include smoke pellets (you had to light them with a match) and flash bombs (well, baked bean cans with 'flash powder' fired using fuse wire and a big battery). I somehow survived this and progressed to using amplifiers with several channels, which meant you could use more than one microphone. Monitors, in the shape of more speakers, were just added on to the main system but pointing back at the band (feedback city!) and yet more lights and fireworks appeared as time went by.

If you did these things often enough without killing anyone or destroying any venues then eventually you'd get noticed by other bands and end up doing a bit of work for them. At this point I started charging a couple of beers per night, there wasn't too much competition as everyone wanted to be a rock star not a 'roadie'. My apprenticeship started with no knowledge at all but I learned something (and still do) every time I went on a gig.

The long and winding road

These experiences somehow led me to my first professional gig with a band that earned its living from playing and were permanently on tour. The gear, distances, and sound all got much bigger rather quickly. My knowledge of pyrotechnics was welcomed but the size of the flashes and bangs would of course have to increase!

The distances can become much bigger when you get out on the road

I learned such a lot from my years on the road with them because no matter what happened, the motto was 'the show must go on'. There would be weather warnings that you shouldn't travel unless your journey was an emergency and we'd go out. Equipment would break down but we'd still rig something up and do the show. The power in the venue would go off when we switched on but we'd find a way around it. Sometimes the sound-level meters would cut the stage power as we were too loud, so we'd run leads from the cigarette machine sockets to bypass it. Whatever happened we would always do the show.

I wouldn't suggest for a moment that you do any of those things. It was a long time ago and things were very different then (including the law). It does illustrate that a bit of ingenuity can usually be brought into play when things get a little tough. This industry isn't one where you can easily take a day off or postpone a job until another day, so you have to become a 'Swiss Army Knife' with your skills and personality.

We had to be flexible to get the gear up in the ski lifts to 8000 feet in the Swiss Alps

All around the world

Wherever you are in the world you will find gremlins who creep in and throw a monkey wrench into the finely honed production. Some of the biggest shows have had major hitches, a famous one being 'The Wall' concert where close to half a million people saw Roger Waters perform an impromptu tap dance while a power failure was tracked down and sorted out. All the necessary preparations had been made on this very technological production but a simple power problem caused the stage monitoring system to fail rendering the artists unable to perform. All this gets forgotten though when your gig goes wrong, even if it's in front of only fifty people it still feels like a major disaster but you will learn from the experience and hopefully look back without too much trauma. It all seems much less stressful in the bar after the gig!

The song remains the same

The basics of a show are still pretty much the same, only the technology has really changed. As an example we used to use an effects box called the Roland Chorus / Echo, which had knobs, meters, and a tape loop running constantly to create effects that we used on vocals. Nowadays you can get a box that's a third of the size and contains many more effects generated in a tiny chip and it doesn't even need a tape change. We still strive to achieve the same goal, though, of presenting a great performance and making the experience as good

Better-quality gear means higher expectations from the audience, so we have to strive for a high-quality sound

as we possibly can for the audience. With the advance in technology has come an advance in expectation from audiences. When vinyl records were the norm people were used to some background noise with their music. This meant that they didn't really hear the amplifier hiss on live shows as a bit of background noise was normal. In today's digital world of pristine production CDs the audience has a different mindset to measure with, so a noisy amplifier will be noticed.

Us and them

If you're looking toward working in this industry as a crew member, then you've probably heard that getting a job is more down to who you know than what you know. There is some truth in this although it doesn't mean that there is no way in. Most of the time, an experienced crew is needed to run a show, so recommendation and past history count for a lot. A company putting together a tour will use crew that they know can perform on all levels including being able to do it without falling out and trying to kill each other when times get difficult. It really is a team effort.

Crew should be able to perform on all levels but you might not be expected to do the soundcheck like these guys

Another very important aspect is trust. You need to be trustworthy in all respects, particularly if you are working with major artists who need to keep their lives private. One mistake here and you will probably never get another job in the business again. If, however, you are good at your job, a nice person, and are trusted, then your name will more than likely be on the list for the next tour.

Actually getting a gig in the first place will probably mean making contact with lots of potential clients such as bands, solo acts, venues, and PA companies and asking if you can help. Be honest about your abilities and if possible send an update to your contacts occasionally (but don't pester them) with any relevant news to show that you're still interested. Shows often require local crew to load all the gear in, help set it up, and then tear down and load out again after the show has finished. This type of work may not be what you want to do in the long term but it is good start. You'll learn lots about the industry and show that you're willing to work.

I'm often asked what the best qualification for the business is and apart from experience I'd say a driver's license. If you can drive then you will often get a gig over someone who can't and apart from anything else you can

A driving license is quite a useful qualification to have. Do try and keep it clean though

travel to other areas to work and also transport gear if needed. Many small-scale productions use small vehicles and many large-scale productions have vehicles both large and small. If you can drive any of these vehicles then it's another skill that you have to offer most clients. It will also help you get some work driving delivery vans when you have no income between gigs!

A large number of crew people work on a self-employed, freelance basis, which is the ideal for prospective employers as they don't have to employ you full-time and don't have to commit to all the other things necessary to employ someone 'on the books'. If they did then they would have to pay out a lot of money to keep you even when there are no gigs or tours for you to work on. As far as you're concerned you won't have the security of a regular job and will have to ensure you take care of your own expenses such as taxes, healthcare, pension, and so on but you will be able to work for a range of clients and charge whatever they will pay you. The money from a tour can look very good indeed when compared to some other occupations but bear in mind that you'll probably be working for many more hours without proper breaks, you won't be going home for possibly several weeks at a time and at the end of it you'll usually be unemployed for a while. I won't go into the details of freelancing as it is different in every country but there will be plenty of advice available locally.

For artists reading this book, I have spent quite a few years fronting my own band, so I understand your perspective as well as the one from the crew. The benefit of performing experience does help me when working with bands and artists who aren't too technically minded as I understand what they're trying to achieve. Many crew folk are also musicians and it's not unknown for crew to play a support slot for the main band (I know, I've done it).

For crew folks reading this book, I've worked as sound engineer, monitor engineer, backline technician, stage manager, lighting technician and all the associated jobs like driving and struggling with heavy gear on single gigs and tours in widely varying venues in many countries using some of the best and worst gear available.

I was in a band myself for some time, so I know what it's like up there

Author at work

I've been lucky enough to have worked in many places ranging from small bars to legendary venues such as The Cavern, The Royal Albert Hall, Abbey Road as well as St Paul's Cathedral (among many other places of worship), theaters, clubs, outdoor and indoor festivals and every other type of venue. Sometimes it's easy and sometimes it's hard, but I do like a bit of a challenge and still love the life.

View from the 'workbench'

I also run a recording studio when I'm not touring and write articles and reviews for music magazines; it all helps fill in the gaps for a freelancer!

Road to nowhere

So if you want to work in the live music business, start now by forgetting about glamour, luxury and an exotic lifestyle, it's time to find out about the real world including:

- Lack of sleep
- Heavy lifting
- Very long hours

- Intermittent feeding times (if at all)
- Endless journeys
- Weekends. What are they?
- Visiting places that you wouldn't normally see
- Meeting people you wouldn't normally meet
- I gotta sleep where?
- No bathroom (sometimes)
- The tour bus designed to keep you awake between gigs
- Hotel staff designed to wake you up 6 hours early
- Working on fantastic shows
- Eating interesting new foods
- Better job satisfaction than most careers
- Becoming part of a 'road family'
- Driving for 8 hours, doing a gig and then driving back again
- Realizing that 'This Is Spinal Tap' probably is a documentary, not a comedy
- Finding ways of filling hours of boredom without it costing anything
- Not having to work 9 to 5
- Getting very lost indeed
- Staying in some exotic hotels (sometimes)
- Driving on the wrong side of the road
- Playing with some great gear
- Foreign healthcare (and what it costs)
- Interesting foreign laws you've just broken
- Great chill-out time on the bus
- How to make things with Gaffer tape
- How to take excess baggage for the cost of a few albums
- People telling you how lucky you are
- Learning something new every day

Welcome to the world of rock 'n' roll. Have fun!

2

We are family

People are what live shows are all about. Without them there wouldn't be much point. Apart from the audience, there can be any number of people responsible for putting the show together. The artists will be the primary focus and may even control the whole of their show (a solo vocalist may have their own sound and lighting system operated from the stage), but most shows will rely on other people to control many aspects for the artists. The size of the show will often dictate the number of crew required. Depending on the venue, they may be crew who work at the venue, visiting crew, or a combination of both. If they are from the venue they will often be called 'staff', but I'll use the term 'crew' for all as there is often common ground and most crew could be from either stable. As an example one band may have no crew of their own and use all venue crew whereas another band may bring some or all of their own crew.

Large shows will require the same crew as small shows plus additional people to perform other jobs and often managers to supervise the extra people and logistics required. A small touring show may have just a single sound engineer as crew, but a festival may have many visiting engineers employed by visiting acts as well as one or more engineers who look after the system and mix the sound for acts that don't have their own engineer. The festival will also have lots of other crew and supervisory staff.

We'll take a look at some of the more common crew members, their place in the overall scheme of things, and what their responsibilities are. This will mean touching on many subjects that you may not know too much about, but we have to start somewhere. You can always come back here or consult the Glossary.

The sound engineer

Most acts working on a professional (or even semiprofessional) basis will employ a sound engineer to look after their overall sound requirements. Duties are mainly to provide a good quality and level of sound for the audience that is mixed well so that all members of the act can be heard properly along with any material that has been prerecorded such as backing tracks and sound effects.

If there are no other sound crew the engineer may also provide monitor mixes to the stage from the Front of House (FOH) mixer. Although not ideal it is possible and often done, provided the mixing console has the correct facilities.

He/she may also be the person who sets up all of the microphones in the correct places. In fact, they may do almost everything if they are the only crew member!

The sound engineer may specify the equipment required for the show either by sending advance documentation to venues (see Chapter 5) or by arranging for it to be taken on the tour in conjunction with whoever is providing gear for the tour.

One of the first to arrive, on some shows, the sound engineer may specify where speakers are to be placed and will liaise with venue crew if it is deemed necessary to provide a feed into the 'house' system. This is sometimes used to provide coverage in awkward areas such as balconies where the venue may already have speakers placed but the touring rig may not easily be able to cover. Many engineers will also carry their own 'FOH rack' with their own set of effects and processors that they are familiar with. These can be used with a venue's own mixer to maintain a good level of consistency in the sound from one venue to another.

Sound engineers always look busy, it's almost as important as being able to mix the band

The backline technician

If only one backline tech is employed then they are pretty much in charge of the stage unless they are a specific type such as a guitar tech. Responsibilities include setting up and running of the backline (onstage amplifiers and equipment) and instruments. This is basically what the act uses and includes things such as drums, bass and guitar amplifiers, props, sometimes stage clothes, special effects, playback machines, and so on. The backline tech will usually arrive with the sound engineer and put all the stage-side components of the show together. They work with the sound engineer to make sure it is all going out to the Public Address (PA) system as it should.

He/she will also keep everything in good order and change strings, skins, and so on as necessary and repair or replace worn or broken items.

You'll sometimes see backline techs swapping guitars with band members between songs and trying to run on and fix things without being seen during show mishaps (this is why they usually wear black). They are also quite often seen doing 'line checks' before the band start by playing the various instruments (often the only bit of fame they can get). Backline techs may also be responsible for more

Restringing before the show. A technician's life can be a busy one

mundane tasks such as putting out the drinks and towels around the stage and locating batteries, strings, Bourbon, and so on in the middle of nowhere when all the stores are closed (sometimes they even do this for the band!).

Many of the bigger name acts will have numerous techs, usually one for each artist. Then their sole responsibility is to keep their artist happy when onstage.

Guitars ready for the show

The monitor engineer

Not a job that is usually volunteered for as it is arguably the most difficult job going. It is often the job with the most complaints and the least amount of thanks from artists. The audience never hear the monitors, so they won't tell you how good they sound and the band will only tell you (read: scream at you) when they aren't hearing what they want.

Monitor mixing is like mixing the band but several, or many, times over

You do get to see the back of the band's heads though

The monitor engineer may have to create many sound mixes to suit the requirements of each performing artist on the stage, so the more artists there are the more the number of monitor mixes that may be needed. Each artist will probably want to hear different instruments to the other artists. The monitor engineer has to provide all of them and maybe a couple of stacks of speakers on either side of the stage as well (sidefills). A six-piece band could have seven different mixes or maybe more if some of them use in-ear monitors as well as speakers.

Monitors are essential for a good performance

When you start putting microphones and speakers in the same space you often get feedback. Imagine the headaches caused by our six-piece that may have twenty or more mics and almost as many speakers pointing *into* the stage. It's OK though as the monitor guy/girl will usually have a rack of graphic equalizers (one for each mix) that have 31 controls each for cutting or boosting frequencies and a mixing console with many, many more controls.

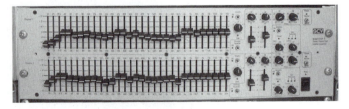

A stereo (or two-channel) graphic equalizer, very useful for monitors

If you get to work on a gig and see the monitors being 'rung out' you'll hear the monitor dude making lots of strange noises (huh, hah, eee, ooo, aaa, tch, sss, one, two) and walking around the stage like a wind-up toy that keeps going back to the mixer before repeating the strange noises and walk. This is a ritualistic per-formance meant to appease the gods of feedback and ask them to not send the dreaded howling down during the show. What it cannot do, however, is prevent the demonic artists and their cunning methods from using their well-practiced techniques of invoking the feedback monsters by the shaking of microphones into speakers and the loud invocations of 'more vocals'.

The lighting engineer

Often called a 'lampie', the lighting engineer is responsible for all things visual (unless there are other people such as video operators) including the setup, upkeep, repair, and running of the lighting (and possibly video and pyrotechnic systems) and often the design as well. If you have designed the rig then you're promoted to the heights of

Keep 'em visible!

lighting designer (LD). The 'lampie' is often the first in if a big rig is to be installed. They will also often be the last out unless there are crew who do the installation and removal as lights often have to be installed above the stage, which has to take place while the stage is clear (i.e. before the rest of the gear comes in).

The lighting engineer often causes a little skirmish by turning off the general work lighting (workers) in order to position, or focus, the stage lighting. This usually occurs when the rest of the crew are doing something very difficult under any circumstances, so when the lampie plunges the stage into total darkness he/she is usually greeted with polite requests to turn the lights back on, as you can imagine.

The lighting engineer may signal other crew such as spotlight (Followspot) operators using a communications headset (comms).

Usually positioned next to the sound engineer with a smaller console the lampie can make or break a show for the band as well as the audience (try playing a guitar solo under just a strobe). Some lampies believe that bands often spoil a good lightshow by getting in the way and some artists complain that the hot, bright lights are pointing at them, making their life uncomfortable, so it's all fair really.

Smaller console, same responsibilities

The tour manager

The tour manager lives in the production office. He/she fulfills a multitude of roles and generally oversees the smooth running of the whole production. He/she is usually involved right from the inception of the tour and will call many of the shots with regard to the show, personnel, and equipment. Usually a key figure in the company that supplies equipment for and organizes the tour, he/she will often have spent many years working in the industry. This means he/she will (or should) know all aspects involved and be able to troubleshoot any situation with ease. Tour managers usually travel with the show and will liaise with the artist, crew, and venue personnel to ensure everything works and everyone is happy. If anyone has a problem the tour manager is the person to see whether you need a shoulder to cry on or you want to rip someone's face off. Speak to the tour manager and all will be miraculously sorted out. Maybe not how you expected but it will be sorted out nonetheless. Cynical crew have been heard to call them the 'Tour Damager'.

Magical creatures of the night, they have the ability to make things (and people) appear and disappear at will,

Happy tour manager, must be a sellout

including themselves. They don't need to sleep or eat like normal folk and if outnumbered have the uncanny skill of being able to talk themselves out of the gravest of situations. Rumors exist that they've been known to buy a round of drinks in the bar at the end of the tour but no concrete proof has emerged. If you're on the crew then they are usually your boss, often sign the paychecks, and decide who will work on the next tour, so I like them.

The stage manager

Usually present when more than one act is performing on a show (such as a festival) to try and keep some semblance of order. The stage manager is totally in control and will tell you so. If the stage clock is 10 minutes fast then no amount of arguing will prove that your watch is right and the clock is wrong. You're on his/her time now and don't you forget it.

If the crew/act have done their part and sent through a full specification of their requirements (a rider) then the stage manager is your best friend and will have all that you need to do the show, except any spare time. They'll tell you when and where you can set up, and provide the gear and stage hands to enable you to do it. If the event has soundchecks you'll be given a fixed time to do it in and when it's your time to go onstage, they'll make sure all the necessary checks are carried out. They'll make sure you are positioned correctly on the stage and keep the show running smoothly. They will advise you of the time you have to leave the stage before the show. Once you reach the end of your allotted time they will haul you off the stage in a very short time indeed, sometimes regardless of if the band is still playing. This is all completely necessary in order for the event to run within its time constraints, which are sometimes strictly laid down by the local authorities or law agencies. The future of the event may depend on it. If they let one act run over then all the other acts will be late and they'll all want an extra 5 minutes. I sometimes have to fend off stage managers when an act I'm working for runs over, but if I'm stage managing an event I turn into a dictator, the same as them!

There are many other titles such as 'Crewchief', 'Supervisor', 'Production Manager', and the like especially on larger shows. They'll usually let you know who they are and their position in the grand scheme of things but if in doubt, ask.

When there are a lot of bands with tons of gear, someone has to take charge

Loving the alien

Working with a good bunch of people can be a very rewarding experience. I often feel a strong family type of bonding develops over time with some crews. Everyone depends on everyone else to some extent and the end result depends on each member doing his/her job properly. Often problems will arise and the crew will pool resources to overcome them, sometimes going to extraordinary lengths. One person struggling with a job will usually find help being offered quickly and the general feeling is one of positive support and resourcefulness from all concerned. As mentioned earlier there is no 'putting things off' until later.

Unlike most other occupations, life on the road brings its own set of rules and situations that are unique. You are often living out of a suitcase in a different hotel every night or driving for hours to get home or living on a tour bus. Each one has its own foibles that are all part and parcel of the job. The people you work with may be great but after a couple of weeks without sleep and sharing a room with them, will their little habits drive you mad? If they keep you awake by snoring or watching TV or they smoke when you don't or they're less than hygienic then it can make the job very stressful and introduce tension. You need to be a diplomat in these circumstances, but you do need to sort something out. If you can't see a practical solution (like earplugs) it's time to talk to them directly, but tactfully, or arrange a little meeting with the tour manager to discuss your options. Don't suffer in silence as the extra pressure will show in your work and you won't be doing anyone any favors by being grouchy. Most people will either know or accept that they snore or have irritating habits but may not realize it bothers you until you mention it. It's also possible that things may be the other way around, so do be prepared to compromise if necessary.

Living this close for a few weeks means you'll have to be good friends

Home

If you're away for any length of time it's easy to get caught up in what you're doing and lose focus of the bigger picture, your life, a big part of which is for most people their home and family. Although you'll probably feel homesick sometimes you'll also probably forget about home sometimes (even though you might not like to admit it), not because you don't care but because you are essentially living a different life and your mind and body will adapt to this over time, making it feel normal instead of your 'other life' feeling normal.

When you know you're going away for a while, particularly if you are going abroad, then make a plan in advance about keeping up contact with home and family, they will miss you as much as you miss them. Nowadays communication isn't much of a problem with mobile phones and the Internet but you need to be careful here as you can run up very large bills. If you end up paying out a lot of what you earned, making the sacrifice of being away and earning little money after paying the phone bill seems a little odd, especially to other people. It's often possible to buy prepaid phone cards that make

Home can seem far away when you're touring, so keep in touch

international calls more reasonable or your mobile phone company may offer special deals. Look into them before leaving. An Internet-enabled laptop or PDA is another option if you can get a cheap connection, possibly using WiFi. If you're staying in hotels then it can be economical for people to call you in your room if you send (or leave) them the hotel number.

If you can (and I'm terrible at this), make a note of friends' birthdays and anniversaries and give them a quick call or text message when you're away. I also like to take home the occasional memento from a foreign place as a future family heirloom, usually something typical of the country I've been in.

It's nice to take a few presents back for loved ones

I take photos so that when I'm sitting in my rocking chair with my pipe and slippers I'll be able to look back at what I've done and where I've been, it might not seem important now but it will, sooner than you think. If I've done a long stretch with a band or company I sometimes put all of the relevant pictures on disk and send them to the band as a gift. It's a nice friendly gesture and good PR for the future.

Contact

Address Book...   Ctrl+Shift+B

Always try and keep in touch with your contacts or at least keep a note of them. You never know when you might be asked to recommend someone for a job and there will also be times when you need to go looking for work. This industry runs on contacts and recommendations to a large extent, so it could make a big difference to your career prospects. I also keep details of many companies that could be useful when out on the road such as PA companies and major manufacturers. Have some business cards with you at all times with your phone numbers and e-mail address. Try and keep the same numbers and addresses as it may be some time before you get a call, but if you've changed your number you'll lose the gig.

I genuinely try to develop a good relationship with my colleagues and peers. Most of the time it is a reciprocal situation as folks don't last long if they are difficult. In general there are a lot of good people in the industry. I have found, however, that you need to be hardworking, conscientious, honest, trustworthy and easy to live with as well as a nice person.

Rumors

One final thing about being on the road is a kind of unwritten rule. 'What happens on the road stays on the road'. This means that sometimes people do crazy and out-of-character things while touring, but you should never discuss them anywhere else, ever. If you destroy the trust then you may lose your career, but it's easy to destroy someone else's career, which may have even more serious repercussions for you. It's simply not worth it.

If you are ever approached by the press about an artist it's better to say nothing, as even a positive comment can be made to sound negative.

3

Ready to rock, but is your gear?

The word 'professional' is thrown around freely nowadays and it seems to have lost much of its meaning somewhere along the way. To me it means someone that's doing a full-time paid job or, more specifically, with regard to equipment, something that's capable of performing to a standard that is acceptable to a professional person in a professional environment. As an example, a cheap plastic microphone that costs less than a jar of coffee isn't what I'd call professional. A tried and tested roadworthy model used by many artists is.

Which one would you trust on the road?

Some readers, especially artists, will have at least one or two items of equipment that they use for gigging, but there's a difference between bedroom or garage gear and touring gear. You don't necessarily need to have the most expensive piece of equipment available. In fact, I tend to take fairly inexpensive gear on the road as it's always a target for theft, and if it gets trashed it isn't such a big loss as loss of expensive gear would be. Watch the baggage handlers at an airport and consider whether you should take an effect costing thousands or hundreds on the road.

Solid rock

The main thing you should aim for is sturdiness and proper casing. Some pieces of equipment just aren't good at traveling unless protected properly. Although it may seem like a boring purchase you should invest in some proper 'flight cases' for your gear whether it be a rack full of effects, a mandolin or a toolbox. There are different levels of protection. If you're just going to be doing small-scale gigs then you can use 'semi-flight' type cases that will handle everyday knocks and scrapes. If, however, you're going to be doing some serious touring then you need some seriously heavy-duty 'full-flight' cases that will withstand most things, even baggage handlers!

(a)

(b)

Lightweight cases (a) are fine for smaller-scale tours but gear like this (b) needs special care and protection

There are many specialized companies that manufacture off-the-shelf and custom cases. They can be found on the Internet or in music trade journals. Proper cases aren't cheap, but consider the cost of replacing your gear and it doesn't seem too pricey. You can also sell your cases later, and they usually have a reasonable resale value.

Always be sure that you make permanent marking of your name and contact details on anything that you take out, but be a little careful not to display sensitive phone numbers (such as artist's) on the outside of cases. Expensive equipment may be best if it is engraved as this is not easy to erase, but permanent marker pen and marking in less obvious places as well as obvious ones can help in the case of a dispute or theft. If you can, make things easily identifiable by using colored straps or tape around handles. This makes identification quick and easy at baggage collection or on a crowded stage.

If you work on a large event with other acts you'll be surprised at the amount of people who have very similar cases.

With large shows you really need to label your cases

High-voltage rock and roll

Safety should be a prime concern for everyone involved with a show, particularly where electricity is concerned. Ensure that *any* electrical gear you have has been maintained well and was safety-tested recently. Some venues will insist on this before they allow it into the building, so always check that you are up to date.

Carry safety electrical circuit breaker devices for your own equipment. Do not use anything that is even slightly suspect. Most venues and rental companies will provide safety devices on their power feeds, but don't take any chances. If in doubt, ask an electrician.

Another thing to check is that you have the correct fuse ratings in every device and carry a good supply of spare fuses wherever you go. If a piece of equipment keeps blowing fuses then it's faulty and should be checked by a qualified engineer.

Depending on where you come from, where you are going to, and what you are taking with you, you might find some anomalies with different power ratings and connectors for mains electricity between different countries. If you are from Japan or North America and you tour Europe then you can't just change the plug. The European voltage is double that of your own, so your gear might just provide an impressive fireworks display as it dies.

Large amounts of power need large amounts of respect. This is the control panel for the power in a theater, obviously lots of electricity present

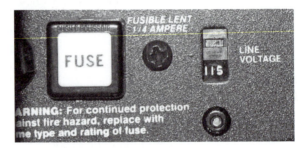

Equipment with a voltage selector switch built-in

Your equipment may have a voltage selector switch built-in, or you will need either a power transformer or a converter. Transformers are required for electronic gear, but you must get the correct wattage rating for all the gear that will be plugged into it plus an extra 25% or so to allow a buffer. The wattage rating for your gear is usually on a plate or label fixed to it close to the mains lead. Again, if you're not sure ask an electrician. If you're touring on this level it's probably best to rent gear in the country you are visiting and just take instruments, tools, and so on. I have known bands that have even purchased gear in another country for a tour and then sold it back to the supplier at the end of the tour. If looked after properly this could be cheaper than renting but don't quote me on that.

If you use a common piece of equipment it's highly likely that you will be able to rent one in most countries of the world (check with the manufacturer). This will mean that you will be instantly familiar with it and also may be able to load your own settings in from a backup medium such as a memory card or laptop.

Sometimes things don't go according to plan with rental gear; an old favorite is the piano sustain pedal that is the wrong polarity for the keyboard. Provided you have enough time, most things can be sorted out. If in doubt take your own pedal!

Batteries not included

Working on gigs at any level requires a certain amount of preparation, that's what this book is all about. When you move up a little the requirements change somewhat, and you need to be looking at the show as a business. Back-ups and spares should be the order of the day and a good supply of consumables, enough to last and then some. Musicians should have backup instruments if possible, and it's a good idea to have at least one spare combination amplifier (a speaker cabinet with amp built-in) that will work for any member in an emergency. There should also be a good supply of skins, sticks, strings, reeds, batteries and so on, for anyone that needs them. Batteries should be replaced well before they start running down as they will always fail in the middle of a show, guaranteed. Professional acts will change them every show regardless.

Crew may also need batteries and should have a good supply of gaffer tape (including both black and white) for a multitude of jobs, electrical tape for repairs, securing coiled cables after use and labeling things as well as cable ties and so on for many uses.

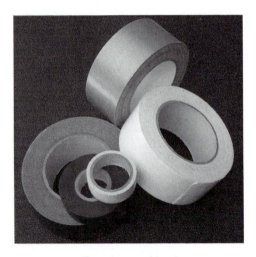

Tape is your friend

It's a good idea to carry a couple of spare microphones and maybe an emergency Direct Injection/Insertion (DI) box, but don't tell anyone that you have them unless it's a real emergency. If they know you have them then they won't provide their own.

Here are some things I usually take on the road with me. I know people who take virtually nothing and others who take much more:

Marker pens both thick and thin.

A decent piece of rag is good for cleaning things as is a paintbrush for getting dust out of nooks and crannies.

Sturdy gloves are a good investment for hauling gear around, especially in colder climates.

Toolkit includes:

A small first aid kit, headache/hangover and indigestion pills.

A notebook and pen.

A comfortable lanyard for hanging passes on.
A Maglite style torch and spare bulbs.
A Leatherman or Gerber type multitool.
A bottle opener.
Soldering gear.
Wire strippers / cutters.
Craft knife.
Fuses (wide range of types).
Gaffer tape.
Electrical tape.
Batteries.
Cable ties.
Spare connectors (most useful types).
Drum keys.
String winder.
Electrical spray cleaner.
Screwdrivers (large and small).
Pliers.
Allen wrenches.
Sticky labels.
Adapter plugs (audio).
Mic clip.
Clamp-on mic bracket.
Cable tester.
Multitest meter.
Battery tester.
Mains tester.
Tuner.
Foam earplugs.
File with rider, specs and useful wiring diagrams.
Business cards.

I often carry drumsticks, strings, a hi-hat clutch, capo, walkie-talkies, backup media, minidisks, and so on depending on the gigs. This all fits into a photographer's style aluminum case that can be taken almost anywhere, although it can't be taken as hand luggage on aircraft due to the contents.

When I'm running the sound I also carry a case with a good set of headphones, a notepad and pen, white electrical tape and marker pens (to label mixers), a CD of music used to test the PA sound, a microphone and lead for talkback to stage, various adapters for my rack gear,

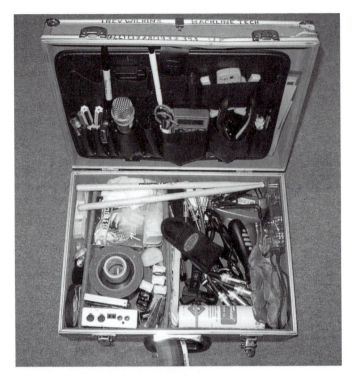

The toolbox…

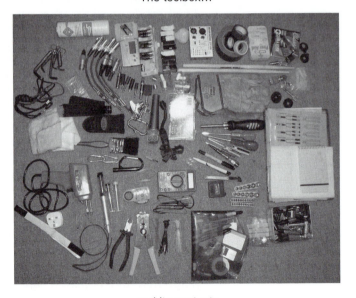

and its contents

an audio interface for my laptop, batteries, a set list and copies of the technical rider.

Advance documentation should always be carried and if possible copies should be held by other members of the team, just in case one set goes missing. Contact details for anyone involved, including venues, promoters and personnel working on the tour, should be easily accessible but only to yourselves. It's also a good idea to get together a collection of useful numbers such as PA and lighting rental companies, music stores and major manufacturers in the industry. These can often help out when things go wrong and you need some assistance far from home.

Depending on the job I'm doing I also carry useful backup information for those times when my brain stops functioning correctly and the old memory cells fail. These include basic wiring diagrams for various cables and tunings for guitars. I should know these things by now, but it's often the simplest, most familiar things that get forgotten easily.

A mobile phone is essential today and a spare isn't a bad idea when on a long tour. I always save sensitive phone numbers under a different name so that if someone gets hold of my phone they won't easily spot any artist details on there. I just change the name a bit so that it isn't obvious, so Trev Wilkins may become Travis Williams.

I, like many crew folk, carry a laptop, which is useful for many things. I used it to write this book while touring! It's also useful as a mobile office, spectrum analyzer, recording system, photographic workshop and e-mail/Internet system among other things. It's passworded so that only I can access it and I use one of those security cables whenever I leave it anywhere. When boredom strikes it can be used to play music and DVDs or to play games as well as for running loads of other software.

Many pieces of hardware (including digital mixers and processors) now have software interfaces, making it possible to tweak your show from a hotel room. Back at the gig you can easily update the hardware when setting up and also retain backups of every setup used.

A good book is a nice escape. When working in foreign countries don't forget that you may not find anything to read in your own language, so buy your magazines before you go.

If you have specific dietary requirements you may also like to carry some supplies for when you can't get the foods you need.

If space allows I use a 'belt and suspenders' approach and carry spare gear that I probably won't need but on occasion it can be a real lifesaver.

4

Hello – Rehearsals and meetings

Planning gigs and tours with some or all of the other people involved is an essential part of the preparations and can help to iron out problems that might not be seen by a single person. Rehearsals are not just for artists. They should also involve the crew so that everyone on the production knows what is going on and gets a feel for the overall show. As well as rehearsals it's a good idea to have meetings to discuss the plans, problems and ideas in depth. It's much easier than trying to talk while someone is practicing a solo.

Getting it right

Before a show goes out on the road it should be well rehearsed so that everyone knows exactly what is going on and when everything occurs during the show. The artists will probably spend some time getting all of their material together and making sure it works properly. If the show is of any reasonable size then there should be some time spent on production rehearsals in order that the crew and artists can sort out various cues for lighting and sound as well as familiarize themselves with the stage setup, and of course each other.

It's a good idea to make notes as things settle down, which can then be copied and distributed to all concerned.

The chances are that much of the show has already been sorted out at least to a reasonable level before production rehearsals, which makes life easier and also costs a lot less in days and wages spent on full production setups. If the crew has a good idea of the show before

Planning, however you do it, is essential

rehearsals then they will have chance to look through and possibly see any pitfalls well in advance. These problems can then be corrected before too much money is spent.

A meeting between key members of the show before production rehearsals can help to sort out such things while general suggestions and ideas can be aired to see if any part can be improved. Whoever is in charge of production should be able to put together a specification for the show early on and have a good idea of what is required, from the total power needed to how many rolls of tape.

Stage, sound and lighting should all be planned well in advance

Who are you?

Depending on your place in the scheme of things you will have certain responsibilities that may require preparation and liaising with specific people well before the show hits the road.

Here are a few pointers:

Sound engineer

Are there any specific requirements from the artists for their overall sound such as 'not too loud' or 'a wall of sound'?

Do any of the artists want special equipment used on their sound such as valve pre-amps?

Ask for a set list (if it changes then ask for an updated one every night).

What cues are there for solos?

Are there any special requests for effects such as delay or reverb?

Try and listen to the act live and recorded to get a feel for what they are trying to portray.

Do they have an intro or outro that you need to play?

Are you required to provide background/walk-in music to be played before and after the show?

Keep your cue sheets handy

If you are mixing monitors from your mixer then you need a list of who wants what in their monitor.
Should you take any spares such as mics?
Will you be expected to record the show from the mixer?
Do you have some music that you can use to set up the PA system that you are familiar with?

Backline technician

You need a set list for every gig and updated ones when it changes (liaise with engineers).

Liaise with the artists and obtain a good supply (with plenty of spares) of any consumables such as strings, sticks, reeds, batteries, and so on that they require.
Make a list of cues for guitar and instrument changes you'll have to make during the show on your set list.
Are there any unusual tunings required?
Will you be responsible for putting out towels and drinks on the stage?
Do you need to call the band to the stage and direct them onto it (in the dark they may need you to show them to their marks)?
Should you have any spares such as cables?

Do you have everything ready?

Monitor engineer

You need a current set list every show.

You should have a list of monitor requirements for each artist updated if necessary.

If the mixes change during the show then it's a good idea to have a list of cues written down.

You need to communicate with the FOH engineer and ascertain which of you is going to provide the phantom power required on the stage.

If possible learn all of the artists' names!

Do they have the right mixes in their wedges?

Lighting engineer

A current set list is needed.

You should try and find out who likes or dislikes smoke and strong lighting, there's always at least one.

You should create a good list of cues or 'looks' to fit the show.

Liaise with other crew as they may need ambient light or darkness to perform some of their jobs properly.

Warn of any potential dangers such as hot lighting on the stage floor that could damage gear or wiring.

Pens, paper and white tape help keep track of lighting cues

Everyone

Responsibilities have to be made clear in order for every-thing to run smoothly. This should be delegated as nec-essary and checked by someone where possible. If the band goes onstage and doesn't have anything to drink they won't be very happy!

Transport and accommodation arrangements should be sorted out and put in writing for traveling to and from gigs and days off when touring. It's sensible for people to travel together where possible.

A list of necessary contacts should be made and pro-vided to all who need it. If there is a tour manager then he/she will have everyone's number. You need to be able to communicate with people that are connected to you, so if you're a drum tech then you should have the drum-mer's number in case you need to talk to him/her when he/she is not there, such as during setup.

Sometimes this can be awkward if artists don't want to give their number out for reasons of security but usually when you work at this level they'll have people who help them and whom you can call, or the tour manager will sort things out for you. In any case always put their name in your phone or documents so that it isn't too obvious such as Trev Wilkins could become Tom Williams. That way if you lose your phone they won't get pestered with calls and find their number is the latest news on the Internet.

If you have close connections with industry, companies try and keep contact numbers at hand for them. This would include manufacturers of high-end gear such as mixing consoles, digital processing and any programmable gear. I often download manuals and information to my laptop for easy access too.

Just in case, it's also a good idea to keep numbers for backup crew and artists.

The above suggestions are just a start and should be discussed during meetings and updated as necessary. You don't have to make a big fuss about changes, meetings often take place on tour buses as a chat about the last gig but it is good for everyone to keep in touch and constantly try to improve things when possible for the good of all.

The lounge of the tour bus, a good place for an update meeting

Production rehearsals

Where possible, even for small productions, at least one production rehearsal should be carried out where everything is set up as it will be on the show. Depending on what you will be taking out on the road this could be a relatively small setup or a very large one. If you're going to use in-house PA, monitoring and lights then you won't need much of a production run beforehand but you may still find that setting up your effects rack (if you are taking one) will be a good move. Running through any intro tracks or playback (backing) tracks helps cement things into your memory.

If you're doing full production for a show you may need several days rehearsing and fine-tuning every part of it. Cues will need to be programmed into any digital gear such as mixing and lighting consoles as well as lists drawn up for other changes. There may also be some last-minute modifications made before the show goes out. The mon-itoring should be well sorted out, which is a major step forward as it will save a lot of time later. Lighting should reach the stage when it is fully programmed and ready to go apart from minor adjustments to suit any differ-ences in venue. As with most things, the preparation is all-important, so full attention should be given to advance planning in all areas.

Sometimes it isn't possible to get much done in advance such as when you're called in at the last minute to replace someone. If this happens then use all the resources you can and ask for information such as tech specs and digital photographs to be e-mailed to you as soon as possible. Any preparation is beneficial.

It's the preparation that makes the show

5

Arranging the gigs

Most of the practical work should be carried out during meetings and rehearsals well before the actual gigs take place. If you're not in control of the whole environment of your shows then there is a certain amount of forward planning that needs to be done so that venues that will host the show are fully prepared for you before you arrive. There are myriad scenarios that you could be involved in, so I'll focus on some of the standard information that should be made available, if necessary, to the venue. If you're carrying a full stage and all of the equipment then you will probably only have to ensure that you have enough space and power to run it all.

Wait a minute Mr Postman

Any information that the venue requires should be sent in advance in hard copy printed form, preferably attached to any contracts. This may not always get to the right people, so if you know who *should* see it then make sure it goes to them and check if they've received it. I've often been told 'we never got your spec'. Sometimes this is used as an excuse for not fulfilling it. Sometimes a promoter has failed to pass on the relevant information to the company supplying the PA/lights as they are trying to save a few bucks by not meeting the spec. Either way, it is a complete pain, so if you call and they haven't received it you can send them another one or point them to a web site that has an easily accessible version of it that they can download. Always carry a few spare copies to hand out to local crew who need them.

One louder

Your specifications should include anything that you think may be relevant to the show that the rental company, venue or its staff need to know about. If you're running a small show then the spec doesn't need to be complicated. If it's a big show then you might need to go into quite a lot of detail including things like:

- Size of the stage
- Risers for drums/keyboards, and so on
- Backdrop or bar to hang your own from
- Number of dressing rooms
- Showers (if touring on a bus you'll need them)
- Food and drinks
- Office space
- Access to office facilities
- Communication facilities (phones, walkie-talkies)
- PA specification
- Monitor specification
- Lighting specification
- Local crew required for load-in and out
- Loading in time
- Power required

- Overnight power line for tour bus
- Filming and recording permitted?

Make sure you can get enough power for your gear by specifying your requirements in advance

You should also specify any important items that you are taking of your own that you will not require from the venue such as:

- Mixing console
- PA
- Monitor console and/or rig
- Lighting console and/or rig
- Stage risers, blocks, backdrops and scenery
- What crew members you have with you
- Any specific gear that you will use, such as drum mics, when they are supplying the other mics.
- Special gear such as in-ear monitors when they will be running the monitor system
- Any radio gear and a list of frequencies used if they are fixed

I often carry a spare microphone or two and maybe a DI box but I don't tell them that.

The documents

Your specifications (or 'rider') should comprise several documents providing all of the necessary information in easily understandable sections. The following are some of the essential ones that you should have. Note that *all* documents should have contact numbers, act name and date of show.

Paperwork is your friend, really

Set list

A list of all the material that will be performed during the show, including any possible changes that may occur and encore numbers. Updated when the set is updated. This is very useful for everyone to keep track of the show and write their own cues based on its contents.

Stage plan

This should be a plan view (i.e. from above) of the stage layout in simplified form showing the positions of equipment such as backline amps, drums, monitors and vocal microphones, if necessary with larger distance measurements on areas such as stage size and distance from front of stage to drums. It's also useful to put the performers' names on so that crew can speak to them in friendly terms, but don't add too much detail as it may confuse the overall picture.

Monitor positions should be marked up to match the other documents that have monitor mixes specified for each performer. You can also label positions for DI boxes and the like if it helps.

Technical specification

A list of your overall requirements and general pointers that may be helpful and save someone having to call you with a simple question. This is the one where you hear about people demanding weird things and bowls of candy with all the red ones taken out but I wouldn't advise putting things like that in. I think it's OK to ask for refreshments but be sure that this is in agreement with any contract or you may end up paying for things later!

Channel list

Depending on its size may also include the monitor list. This provides the necessary information of how your mix is set up and includes your subgrouping, monitor grouping and preferred microphones (if you have any preference). You can make this as comprehensive or as simple as you like depending on who needs to see it and what they will need to know. Even if you are the engineer and will be taking your own console it will be helpful if anything changes such as your mixer, or if you have to miss any gigs and someone else has to stand in for you.

Don't add any unnecessary channels as you will look a bit stupid when they've been set up for you and you don't need them. On the other hand, if there's a possibility that you may need more channels (the guitarist just might use his/her acoustic tonight) then put them in the list.

Itinerary

Fondly referred to as 'the book of lies' by crew, this is the list of what's happening, when and where, which everyone lives by when touring. They come in all shapes and sizes and contain a variety of information but are basically a planner or diary of how the tour *should* run.

Generally, they will have a tour overview with the dates and venues, a list of personnel and a page by page diary of each day and its order of events with their planned times. So you may have something like this:

Thursday 1st November
Joe's Bar
Littletown
Biggsville
B1GGUY
Box office: 01234 56789
Travel time: 3 hours
Load-in: 11 a.m.
Soundcheck: 4 p.m.
Soundcheck support: 5 p.m.
Doors: 7.30 p.m.
Support onstage: 8–8.45 p.m.
Band onstage: 9.30–11 p.m.
Band Hotel: The Grand Hyatt, Fun Street, Littletown (across the street)
Crew Hotel: The Pitts, Pittsville, Darktown (3-hour drive away)

There should also be details of any flights, ferries or other transport reservations, and anything to do with the transport and logistics of the tour. There's usually a fine for losing your itinerary and a slap upside the head for asking a question that is answered in it!

You shouldn't miss your plane if your itinerary is correct

Other useful documents could include a backline spec for gigs where you have to fly-out without your own backline and gear has to be rented. A list of personnel and their contact details is useful but may be included in the itinerary. If lighting is to be installed into a venue using existing gantries, trusses and the like then a lighting plan and spec should also be sent.

Venues will often have their own specs posted up on their web sites or available by request, so if you're going to be using any of their gear you might want to check this out in advance. Bear in mind that specs can be out of date and make a note to keep your own up to date.

Here are some example documents you can use as a basic outline for your own.

Example Documents
Examples of the paperwork that should be supplied to venues in advance of shows

OUR BAND SET LIST

<u>Date</u>. . . . .

<u>Set 1</u>

1. Warming up the crew – Intro CD. Guitar solo. Lead vox only
2. Rock n soul – Keys start. Lead vox and BVs. Key solo. Guitar solo.
3. Stormy Wednesday blues – Guitar intro. Lead vox only. Guitar solos.
4. Dark side of the room – Piano intro. Long vox reverb. Lead vox only.
5. Paperback lighter – All-in. Lead vox and BVs. Bass solo.
6. Just the way you were – Piano and lead vox only
7. Born to be mild – All in. Keys solo
8. Court of the crimson thing – All in. sax solo. Keys solo.
9. Stairway to Devon – All acoustic guitars. Lead vox only
10. Lighter shade of hail – Keys intro. Lead vox only.

<u>Set 2</u>

1. Rock around the block – All in. Sax solo
2. Jumpin' Jack's cash – All in. Guitar solo.
3. I will always love food – Keys intro. Big vox reverb. Lead vox only.
4. Superfry guy – All in. Lead vox and BV.
5. Eighteen with a mullet – Piano intro. Lead vox and BV.
6. Runnin' on plenty – Piano intro. Harmonica solo. All vox.
7. New kid in brown – Acoustic guitars and piano. All vox.
8. Reeling in the beers – All in. All vox. Guitar solo.
9. Bloke on the water – All in. lead vox only. Guitar solo.

<u>Encores</u>

1. TV winners – All in. Lead vox only. Guitar solo.
2. Johnny B Bad – All in. Lead vox only. Guitar solo.
3. Black frog – Guitar intro. Lead vox only. Guitar solo.

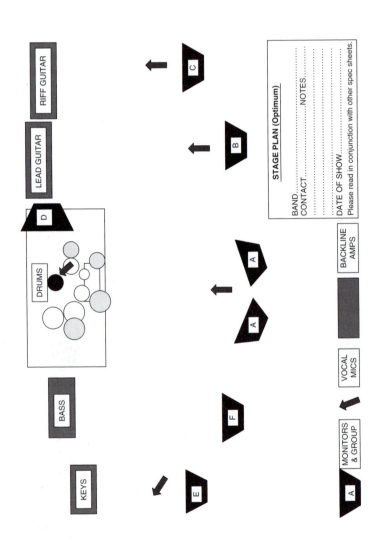

Technical specification

Please read in conjunction with our stage plan and channel list

Band name. . . . .

Date of show. . . . ./. . . . .

Band members

Lead vocal:
Lead guitar:
Bass guitar:
Keyboards:
Drums:

Stage area required is 20 feet (6 meters) wide × 15 feet (4 meters) deep minimum.
If you have a suitable drum riser we will use it.

Sound system should be sufficient power and quality for the venue. A 3-way system of 4 kW should be seen as a minimum.

24-channel front of house mixer is required with 8 subgroups and facilities for 3 effects, preferably returning on stereo channels.

Full 6-way monitor mix will be required driven by your engineer from stage left if possible or from FOH by our engineer if necessary. Full 31-band graphics are required for each mix. We will bring a set of wireless in-ears for the lead vocalist, all other mixes are to be on wedges, pairs if possible. Drumfill should include a sub. We do not require sidefills.

Effects required are 2x reverb, 1x delay with tap facility, all stereo.
Processors required are 10x compressors and 6× gates.
We carry a rack with our own effects in for venues that don't have the relevant facilities.
We don't carry any microphones, stands, cables or DI boxes.

A stereo 31-band graphic is required across the main PA.
A CD player is required in the main PA for our setup and intro.

We need 2 local crew to assist with loading in and out of the venue, suitably experienced people are needed as we have some heavy items.

We also require 12 small screw top bottles of still mineral water during the show for band and crew.

We would appreciate any refreshments that you can provide during our setup as we have probably been on the road for several hours. Tea, coffee, biscuits and sandwiches being especially suitable.

If you have any questions or doubts about this specification please contact:

1. . . . .
2. . . . .

Many thanks!

32-CHANNEL MIXER AND 8-WAY MONITOR LIST

CHANNEL NUMBER	SUBGROUP	INSTRUMENT VOCAL	MIC/DI Minimum	MONITOR GROUP	FX	NOTES
1	1–2	Kick drum	Beta 52	ABCDE		
2	1–2	Snare	SM57	ABCDE	Rev1	
3	1–2	Hats	SM57	BDE	Rev1	
4	1–2	Tom 1	SM57	D	Rev1	
5	1–2	Tom 2	SM57	D	Rev1	
6	1–2	Tom 3 (Floor)	SM57	D	Rev1	
7		O/H L	414		Rev1	
8		O/H R	414		Rev1	
9		Bass	DI Active	ABCDE		
10	3–4	Guitar 1 Riff	SM57	AD		
11	3–4	Guitar 2 Lead	509	D		
12	5–6	Keys L	DI Active	ACD		
13	5–6	Keys R	DI Active	ACD		
14	5–6	Piano	DI Active	ACD	Rev1	
15	7–8	Lead Vox	SM58	ABCDE	All	
16	7–8	Vox 2 Lead Guit	SM58	B	All	
17	7–8	Vox 3 Keys	SM58	E	All	
18	7–8	Vox 4 Riff Guit	SM58	C	All	
19	7–8	Vox 5 Drums	SM58	D	All	
20		FX Reverb				
21		FX Reverb				
22		FX Reverb 2				
23		FX Reverb 2				
24		FX Delay				
25		FX Delay				
26		CD for Intro				
27		CD for Intro				
28						
29						
30						
31						
32						

MONITOR GROUP	PERFORMER/POSITION	NOTES
A	Lead Vox Center	
B	Lead Guitar Center S/L	
C	Riff Guitar S/L	
D	Drums Center	
E	Keys S/R	
F	Bass Center S/R	
G		
H		

BAND.................................... DATE OF SHOW.....................

ENGINEER...

By Trev Wilkins. ©Tight Digital Music.

32-CHANNEL MIXER AND 8-WAY MONITOR LIST

CHANNEL NUMBER	SUBGROUP	INSTRUMENT VOCAL	MIC/DI	MONITOR GROUP	FX	NOTES
1						
2						
3						
4						
5						
6						
7						
8						
9						
10						
11						
12						
13						
14						
15						
16						
17						
18						
19						
20						
21						
22						
23						
24						
25						
26						
27						
28						
29						
30						
31						
32						

MONITOR GROUP	PERFORMER/POSITION	NOTES
A		
B		
C		
D		
E		
F		
G		
H		

BAND.................................... DATE.....................................

ENGINEER...

By Trev Wilkins. ©Tight Digital Music.

The Dream-on Tour itinerary

DATE: Thursday 12th March   TRAVEL TIME: 4 hours
COUNTRY: England   TOWN: Bigsville

Venue

The Big Gig
The High Street
Uptown
ABC 123
Contact
Big Jim: 09090 765 432
bigjim@bigjims.co.net

Load in: 12.00 p.m.
Soundcheck: 3.00 p.m.
Doors: 7.30 p.m.
Stage: 8.30 p.m.
Curfew: 12.00 a.m.
Sets: 2 × 45 minutes

Support: Puppet show. 7.45–8.00 p.m.

Hotel

The Big Grand
The High Street
Uptown
ABC 123

061234 78910
Booking Ref: 0123456/A

Notes

Champagne and caviar will be provided on arrival for all crew.
All hotel rooms will have free bar and Jacuzzi.

6

The knowledge – Basics that you need to know

There are some basics that should be understood before we go into any more detail. You may be tempted to skip this section but do persevere and try to get a grasp of these basics as the rest of the book will fall more easily into place if you do. You do need a strong knowledge base, as others in the field will assume that you know about certain things.

Later chapters will go into more depth in specific areas, so some of the information here only covers the basics. Once you know the basics, each chapter will be easier to understand.

About sound

Without wanting to get too much like a boring textbook I think it's a good idea to have some knowledge about sound. You will undoubtedly come across certain terms that relate to it and if you understand what they mean then you'll be better equipped to deal with them. As you probably know sound consists of waves, which we can understand easily if we show a representation of them like this.

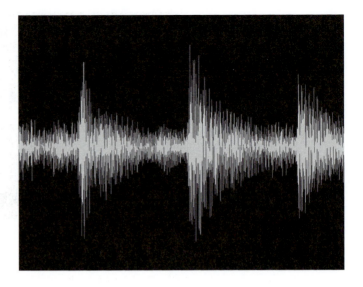

The level (amplitude or pressure) of the sound can be observed by an increase in the height of the wave, which we usually measure in decibels (dB). A decibel is one tenth of a Bel and the term 'Bel' is named after Alexander Graham Bell who invented the telephone. This measurement is a logarithmic measurement, so it doesn't have a fixed value in the same way as an inch or a ton does; it can be used to show the difference (ratio) in level between two sounds though as in 'sound 1 is 10 dB louder than sound 2'. A 'weighting' system is used as a point of common reference, so if we use the standard of 'A' weighting (used for general noise level measurements) this gives us some tangible points that are easy to grasp. Here is a list of sounds and their approximate rating measured in dBA:

0	Barely perceptible sound with normal hearing
30	A soft whisper
60	Normal conversation level
120	Jet engine on runway/rock concert
170	Shotgun

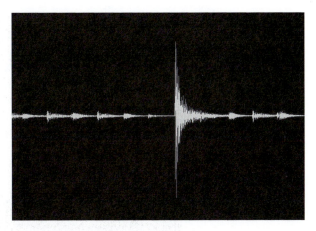

You can see the difference in level between the quiet sounds and loud drum hits in this wave

Prolonged exposure to levels over 85 dB are considered to be dangerous to hearing over time, so you should try and avoid long-term exposure whenever possible.

As a guide to what some terms mean, here is a diagram showing what parts of a wave certain terms refer to.

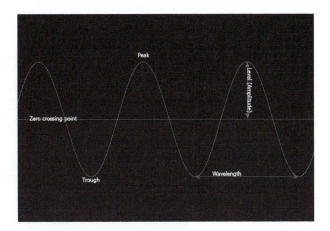

Frequencies

The frequency of peaks and troughs within the sound wave can vary widely and the higher the frequency (more peaks in the same time) the higher the pitch of the sound. We use the frequency measurement a lot in live sound, especially when running monitors. A good monitor engineer will be able to identify, just by hearing a speaker feeding back, which frequency range is causing the problem and decrease the level of that part of the range in the speaker using a graphic equalizer. Many pieces of equipment have a stated range of frequencies (such as 70 Hz to 18 kHz) that they can handle, which dictates their suitability for various jobs, such as microphones and loudspeakers.

If you want an easy guide, think of bass or low-end sounds as being low frequencies and high-pitched or treble sounds as being high frequencies. Bass will usually be in the range measured in hertz but middle and high frequencies will be measured in kilohertz or thousands of hertz. Ten kilohertz is ten thousand hertz, quite high, while 10 Hz is ten hertz and is very low, in fact lower than we can hear, so it is subsonic. We can hear somewhere between 20 Hz and 20 kHz when young and with good hearing, but as we get older our hearing range diminishes, particularly the higher part.

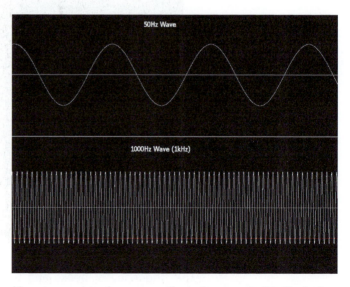

You can see how frequency affects the wave in this illustration

Phase

When referring to 'phase' with regard to sound we generally talk about sound being 'out of phase' as this is a problem that can manifest itself as a noticeable drop in level. If we look at our wave diagram we see that the wave goes above and below the central line called the 'zero crossing point' as there is zero signal, which is in effect silence. If we have two signals coming from the same source (e.g. two mics on a drum, one above and one below) then it's possible that the signal from one may be the opposite (or different) from the other one; when this happens the difference causes a drop in level. If the signals are completely opposite so that the peak of one is at the same time as the trough of the other (out of phase) then the result can be a canceling out of the signal completely. It is also possible for speakers or cables to be wired incorrectly (+ and − connections reversed), which will cause phase problems. There are usually switches on mixers to reverse-phase on incoming signals if necessary but if not then a short adapter cable with reversed polarity wiring in the connectors can be used between the source and the mixer.

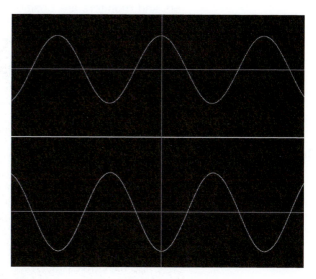

These waves are out of phase and canceling each other out, causing silence. Notice how the peak of the top wave is in the same position as the trough of the lower wave

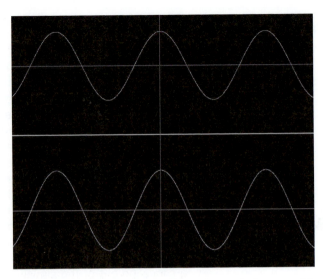

These waves are in phase as their peaks are in line, providing a good strong signal

It is also possible that speakers projecting the same content could be positioned incorrectly, which may cause phase problems. It's important to ensure that PA speakers and monitors are positioned in line with each other. If there's any doubt then just move one backward or forward and see if the sound improves. If so then they were out of phase.

Phase problems can sometimes be heard as a 'swirling' of the sound, which is exactly how the 'phaser' effect, sometimes used on guitar, gets its name.

Sound pressure level

Sound pressure level (SPL) is a term we often find used in equipment specifications to show what level of sound they can take before distorting. It is also a term that is used to inform you of the current noise level that a venue has. This means that you have to keep below this limit or you may find the power cut by a trip device or you could have someone watching over you who keeps telling you to turn down. The best thing here is if they provide you with a SPL meter so that you can keep an eye on the level yourself.

SPL is usually measured in decibels. It's useful to note that a scale called 'weighting' is usually used to measure the levels. There are several different weighting scales but the most common are A, B and C. For our purposes, the C scale is probably the one you'll see most. A weightings are generally used for lower levels up to around 55 dB, B weightings for midrange levels from 55 dB to around 85 dB and C weightings for higher levels. The readings show the change in the current ambient level of sound and depending on the measuring device's settings one can measure average or peak levels.

Mono or stereo?

In the formative years of amplified live sound all the signals in the mix (which often weren't many) were sent down a single signal path that went to the speakers wherever they were placed. If you listened to the left side of the PA you would hear exactly the same as if you listened to the right side. This is called 'mono' or 'monophonic' to give it its correct name. As technology advanced a standard appeared, which allowed for sounds to be sent to either left, right, center or any position between which more closely matches what we hear; we have two ears, so we hear left and right sides slightly differently. This is stereophonic sound (stereo), which has become the normal method for producing both recorded and live sound for many years. It enables mixing to be more defined in a spatial sense as the engineer can 'pan' (short for panoramic control) any sound anywhere from fully left to fully right.

Mono is still used extensively for monitors. Most sound sources are actually monophonic, so the placement is key. However, some sounds are stereo (such as keyboards or stereo mic'd ensembles), so they are treated as a 'stereo pair' and will use two channels panned appropriately. It is also possible and common to apply stereo effects to mono sources, so a stereo reverb may be applied to a vocal to make it sound 'wide' and more ambient. Stereo is the usual system for most larger PA rigs, which requires each side to be driven separately by amplifiers, which usually consist of two separate amps in a single case, one used for left and the other for right. The bass part of the rig may be driven in mono, as low frequencies do not have directional content. The two amps are often combined in a way called 'bridged mono' to provide the necessary extra power required, a good use of resources!

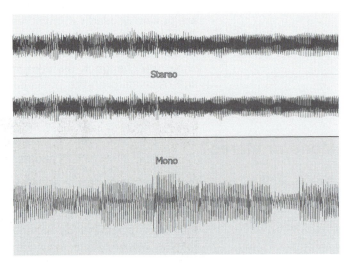

A stereo guitar wave (top two) and a mono bass wave (bottom)

Analog versus digital

The change from analog to digital in the live environment has not been as fast as many other areas due in part to the stability and processing speed required for live sound as well as a resistance from 'old hands' for a variety of reasons, including the lack of hands-on facilities of many products. A sound engineer who has used analog equipment for a long time will find it easy to change the bass level on his/her kick drum while simultaneously adjusting the compressor threshold with the other hand. This may not be possible with some digital devices and could even require a search through menus or pages to get to various settings, which is not intuitive to him/her. They need knobs and faders! Manufacturers who design for the studio first will not be their friend as immediacy isn't their priority. Equipment designed specifically for live use has to take this into account, so recent developments have been more geared to this. Digital technology is much more widely accepted than it was just a short time ago. Units that didn't require fast access were the first to be accepted. A digital speaker management system that is set well before showtime is not a problem; it is the same for effects units, which can have presets created and saved beforehand, which are easy to recall. The move has been gradual but is increasing at a fair pace as many of the initial problems have been overcome. Processors are fast enough to cope and the quality is up to a high-enough level without sounding too 'clinical'.

There are advantages to digital systems. One of the best I've seen recently is the digital snake, which is basically a piece of telephone cable that can carry all of the signals to and from the stage successfully. This is traditionally accomplished by a thick and heavy multicore cable that is difficult to handle and very costly. The new digital cables are so cheap that you can afford to run a spare one alongside the main cable, just in case the main one fails. Large events will probably see the cable being buried before the show and just left in the ground after, as it will be more economical to leave it (although I'm not sure how environmentally friendly this is).

I believe that both analog and digital are valid. If you are flexible enough to be able to use both then you will be more valuable to any prospective employers.

Stage directions

Aside from the unstoppable march of technology there are some things that will always apply as they have worked for many years and still work today. One of these things is the set of stage directions that have been used in theaters, probably since before Shakespeare's time. When working on a stage it's useful to have common terms of reference for where you are on the stage or where items have to be placed, so there is an accepted set of directions you should be familiar with.

Upstage is the rear of the stage, furthest away from the audience. Easy to remember on a raked stage that slopes downward toward the audience.

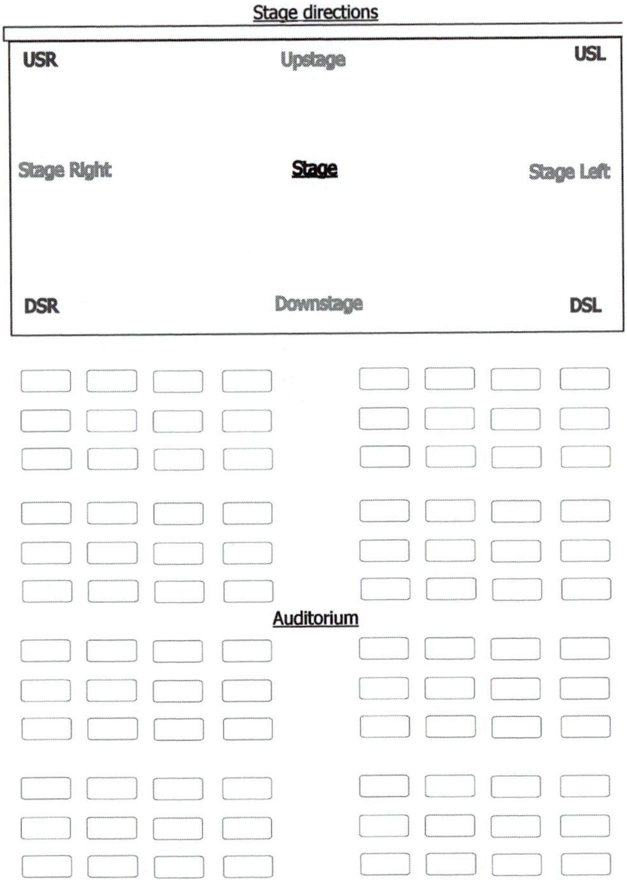

Stage directions

USR	Upstage	USL
Stage Right	**Stage**	Stage Left
DSR	Downstage	DSL

Auditorium

Downstage is the front of the stage nearest the audience.
Stage Left is your left if you are facing the audience.
Stage Right is your right if you are facing the audience.

So Upstage Left is at the back to the audience's right,
Center is obvious. Often cases will be marked with the
stage position of their contents, which may be abbreviated, so USL would be Upstage Left. I have had venue
crew try and test me out on this by asking questions
like 'where does your drummer go upstage left or downstage right?' so I usually get a bit facetious and say 'at
the back'!

The building blocks of a show sound system

Although there can be many variations in the way sound
systems are put together there is a standard used for
medium to large rigs that will be almost universal as it is
a tried and tested method. Smaller systems usually have
less complexity in their components and fewer crew to
operate them, so monitors may be operated by the FOH
sound engineer or there may not be any monitors at all.
Sometimes, an extra pair of speakers carrying the full PA
mix will be used by facing them into the stage so that
the performers can hear everything. This would then only
need a level control and maybe equalization (EQ), which
is easily achieved.

Front of house

The sound engineer should be positioned where he/she
can hear the PA system properly and also be in a reasonably good place, such as the middle of the audience, to
ensure that, in general, the mix is acceptable around the
rest of the venue. During the setting up of the system
and soundcheck, he/she will listen in various areas that
the audience will inhabit to see if there are any anomalies that need to be taken care of. This will usually be
difficult during the show, so a good mix (FOH) position is
essential.

A good mix position is essential

In the mix position will be the mixing console along with racks containing any *outboard* equipment that the engineer needs such as effects and equalization for the main PA system. If a digital console is used then these may be contained within it but it is still possible that external racks will be used if the engineer has favorite 'toys' that he/she prefers. The mixer will usually have a pair of power supplies also in a rack. The reason for using two is that if one fails then there is a spare to take over. Mixers don't usually fail totally. They may lose odd channels or sections but total failure would probably be caused by a power supply failure, hence the spare.

Recent advances in digital technology have seen the introduction of speaker processing units, which will often live in the FOH racks as they allow the engineer to control a range of system parameters that affect its performance and response. Sometimes a 'system tech' will be present when visiting engineers are using a rented or in-house system. They help them achieve their mix very quickly and generally look after the system, which they will have a good knowledge of, for the duration of the event.

Front of house processing racks and effects

The PA system

If you didn't know already then PA stands for Public Address, as this is what it does, addresses the public. It primarily consists of speakers and amplifiers that receive the mix signals, amplify them and project the sound to the audience. The mix signal is sent from the engineers mixing console at FOH but in order for it to be mixed it has to be sent there in all its separate pieces (vocals, drums, brass, guitars, etc.). This is achieved by using boxes on the stage that have all of the signal cables plugged in. They have thick cables that contain multiple cores that carry all of the signals separately to the mixer where they are inserted into the mixer as separate channels. These are generally called 'snakes' due to their appearance or 'multicores' due to their construction. They can also carry signals other than the main mix back to the stage, which are referred to as 'returns'. This is what you would use if mixing monitors from FOH: A 'return' path.

As larger speaker systems consist of banks of speakers that work for a predetermined range of frequencies (such as bass, middle and treble); a controller is used that splits the main audio feeds into the necessary frequency

Stage boxes

ranges and then sends the correct range to the correct bank of speakers. This piece of equipment is referred to as a 'crossover' as it is used to set the point at which the ranges 'cross over' to each other. It is placed between the output of the mixer and the PA power amps so that the audio going to the PA is processed before it is amplified. The amplifiers will be connected to the speakers in banks that correspond to the various frequency ranges. The bass cabinets will have amplifiers driving them, which will receive the bass output from the crossover, the mids will receive the middle frequencies from the crossover and so on. The assignment of amplifiers also works well this way, as different frequency ranges require different amounts of power to drive them (sub-bass speakers will need much more power than high-frequency horn drivers). As an example you may need 1000 W for a bass driver (speaker) and only 100 W for a high-frequency driver. The crossover may be included as part of a digital speaker management system or could be a stand-alone unit. It may be positioned at FOH by the mixer or on the stage end. It is often placed with the power amps to enable easy connection.

Depending on the number of 'banks' or frequency ranges that are used the system will be referred to as 3-way or 4-way so that it is easily understood. A 3-way system would generally be bass, middle and treble whereas a 4-way might have an extra sub-bass content.

In general, the speakers will be arranged with the lower range at the bottom (bass) and the higher ranges above. Although there is no fixed way to do this, two systems are mainly used, namely stacking speakers or 'flying' them. Stacking is pretty obvious while flying means that the cabinets are suspended in the air usually from steel cables and aimed from there using other cables to hold them steady. You can also ground stack some and fly some if it provides the necessary coverage for the audience. This is often used where there are balconies in venues that wouldn't be covered by a complete ground-stacked PA. The popular line-array systems usually have the bass content ground stacked and the rest flown and angled to provide full coverage. Bass frequencies do not contain directional content, meaning you don't need to aim them very accurately as the sound will disperse well anyway. Higher frequencies do contain directional information, so it is crucial that they are aimed properly. An example of this is if you have a surround-sound system at home you'll probably have a single sub that can be placed in the corner of the room and sound fine but the other speakers have to be placed correctly or any positional sounds (such as a car driving by) will sound wrong and displaced. High-end systems may have computer software for 'aiming' the speaker array in a calculated manner to suit the venue and audience.

Some speakers have amplifiers built-in and therefore only require a signal to be sent to them, which can save having amplifier racks. They also need mains power and are obviously a lot heavier than 'passive' speakers, so there is a trade-off.

More on speakers and amplifiers in Chapter 7.

The monitor system

Essentially part of the PA system, this is what the artists have so that they can hear themselves properly. If you've

A speaker 'ground stack'

ever heard someone singing along to their personal stereo with their headphones on then you'll understand how difficult singing is when you can't hear your own voice properly!

Full-fledged monitor systems are like a bunch of little PA systems insofar as they consist of a mixer, amplifiers, speakers and processing gear to make them sound good. The mixes they contain are usually different from the PA mix as each artist will have their own monitor mix sent to their own monitor speakers (or in-ear monitors). In order to do this the stage signals (mics and instruments) are split before they are sent to the FOH mixer and they are also sent to the monitor mixer. This is a special type of console (although some FOH mixers will do the job). It is capable of sending out a number of different mixes, so it is essentially a matrix bringing all of the signals in and being able to send them to any output at any level (see Chapter 9 for more information). The physical requirements of a monitoring system mean that it is a large (and heavy) part of

Speakers 'flown' to cover the balcony seats

a stage setup. It is a complex and time-consuming thing to put together and operate, requiring plenty of sound-check time to get right every show it is used. The monitor console is set up at the side of the stage where a good sight line is possible with all of the artists so that they can signal or ask for any changes they require easily. Monitor speakers are the wedge-shaped boxes that you see on the floor of stages. They are this shape as the speakers are then facing the artist without being in the way of the audience's view too much. Drummers may have a large stack of speakers called a 'drumfill' and often there will be stacks either side of the stage facing inward, which are called 'sidefills'. In-ear monitors may also be used. These can be wired for performers who will not move far, such as drummers, or a wireless radio type for people who do move about a lot.

Monitoring is mixed from the stage

Back in the bar

Obviously, the specification we are talking about here will be much reduced for small venues and amateur bands. But it is the type of system that will benefit any band when used properly as they will hear themselves and the audience will hear a good-quality PA sound. A big rig isn't necessarily going to be there for volume. It's as much about quality, which is often a result of having 'headroom' to spare on a system. If we compare it to trucks then a truck that is designed to do 50 will be close to the limit of its performance when it reaches 50 but one that is designed to do 80 will be performing with ease and just 'ticking over' at 50. So it is with audio. A 3 kW rig may be close to distortion in a small hall when a 10 kW rig will be running just fine. As with most things, quality costs more but it does make a difference. This is why large rental companies will buy amps costing thousands each. They sound great and are totally reliable. Your amp costing a couple of hundred will sound OK with your speakers in a small room but it probably won't last too long if gigged hard. If it has a bit of abuse it might just fail and take the speakers with it. Rental companies can't take that chance.

Recent trends and advances in technology have seen the great big old speaker boxes and amplifiers replaced by much smaller and more manageable units that have higher efficiency and can provide better sound in smaller packages. This is good for everyone concerned, especially anyone who has to carry it.

Let's rock!

7

Speakers and amps

In order to reproduce sound or project it above its normal acoustic levels we need to amplify it and force it in the desired direction. The basics of this involve a speaker (or loudspeaker to give it its full name) and an amplifier, or amp for short. These pieces of equipment are designed to reproduce the audio signals that they receive and project them at a higher level, and/or provide a greater spread of sound in a way as faithful to the input as possible plus any enhancements that are required to make the sound more pleasing. Although we're mainly interested in PA systems here, the use of speakers and amps obviously extends to the amplification of instruments as well. While speakers are similar for PA and instruments the amplifiers used are not quite the same. As the signal is being adjusted by a mixing console and other processing equipment there is no need for the amplifiers to have controls such as EQ, reverb or extra gain stages; we usually call them 'power amps' as that is their main purpose, to power the system.

Without going into a physics lesson we'll take a quick look at the principles behind what we're trying to achieve with our speaker and amp.

Sound radiates from its source as vibrations, which are often likened to ripples on water when a stone has been thrown in. These 'waves' have differing frequencies, which, very simply put, would be the frequency of the peaks of our ripples in the water, counted over a measured time such as a second. Low frequencies could be 50 times a second, for instance, and high frequencies could be 10 000 times a second. Without jumping the gun the low would be the lower end of the audio range that we call 'bass' and the high would be nearer the high end that we call 'treble', just like on your home music system. Most

(a)

(b)

To make your sounds heard you'll need some speakers like these (a) (preferably in cabinets) and some amplifiers like these (b) to drive them

sound sources produce a range of frequencies and so they are likely to consist of a very complex pattern of waves.

We use a measurement unit called hertz to express the frequency of sound, which translates as 'cycles per second' and is abbreviated to 'Hz'. In true human fashion it is named after the scientist Heinrich Rudolf Hertz (1847–1894), who was the first to transmit and receive radio waves (useful to know for quizzes). When we reach

You may find simple EQ controls on home hi-fi equipment

into the thousands of hertz we add the prefix kilo (meaning thousand), so 10 000 Hz would equal 10 kHz, much easier to write and remember.

As examples, middle C (C4) on a piano is 262 Hz, A4 is 440 Hz, low E string on a bass (E1) is 41 Hz and instruments such as cymbals may work in the high kilohertz up to around 16 kHz and above.

You could also go to megahertz (millions of cycles per second) or gigahertz (billions of cycles per second), but

Each note on an instrument has a different frequency

we can't actually hear above around 20 kHz and so we don't need to go there for live sound! The lower end of human hearing is around 20 Hz but low frequencies can often be felt physically as well. Hearing range will weaken over time, so as we grow older the higher figure usually drops to maybe 12–14 kHz. Bear this in mind and do not accelerate it by exposing yourself to high levels for long periods of time.

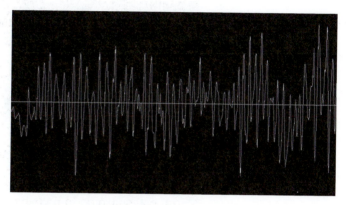

Various sounds added together produce complex waves

Levels below our range of hearing are called 'infrasound' and levels above are called 'ultrasound'. If you can hear infrasound then you're probably an elephant and if you can hear ultrasound you could well be a dolphin or a bat and you may not get a job in this industry.

Our sound source is radiating all these frequencies, which are being picked up by a microphone. This piece of gear absorbs the vibrations and turns them into a small electrical current known as an audio signal, which can be sent down a wire to an amplifier. The amplifier will receive a very small impression of the sound that is too minute to drive a speaker but that is a pretty good representation of the real sound (if it's a good microphone). The amplifier takes this signal and increases it by using a high electrical current to reproduce it at a level sufficient to drive a speaker. Smaller amplifiers called 'pre-amplifiers' (preamps) may first increase the signal and then pass it on to power amplifiers. Mixing consoles contain pre-amps that send a workable signal through their own circuits, then pass it down the line to the power amplifiers. In essence, the amplifier increases the 'amplitude' or level of the signal. We perceive amplitude as how loud something is, so imagine that the microphone is sending a very quiet sound level that the amplifier makes louder.

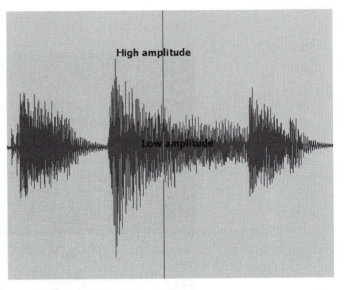

Low amplitudes produce small waves while high amplitudes produce larger waves. Short, high-amplitude peaks are called 'transients'

A speaker is very similar in construction to a microphone (in fact, you can use a speaker as a microphone but we won't go there) as it has the same basic components, namely a diaphragm and a means of translating electrical impulses into vibrations (or the other way round for a microphone). A dynamic microphone, the type used most in live sound, works as the diaphragm vibrates in sympathy with the sound it receives, that is to say that it vibrates at the same frequencies. This vibration is passed on by a physical attachment to a coil, as it moves the coil past a magnet this generates a small electrical current that we can think of as an electrical imprint of the sound. After amplification the speaker uses a coil to receive the incoming signal (a larger version of the sound imprint) and in conjunction with a magnet it translates this into vibrations that are made audible by the use of a cone, usually made of stiff paper or fabric, moving back and forth to move the air at the correct frequencies. The coil is usually attached to the cone, which is flexibly attached to a framework (chassis) that holds the magnet. Speakers are generally circular in shape. Often, larger diameters are used to reproduce lower frequencies and smaller ones for higher frequencies, although this isn't always the case.

The coil windings for a speaker

To produce the full audible range of frequencies at an optimum level a range of speakers is usually used with speakers designed to work at their highest efficiency over a restricted range. Lower-frequency ranges, often referred to as 'sub', require more energy to be produced by a speaker, so it's usual for this area of the audio spectrum to be covered by dedicated speakers or 'subs'. The higher-frequency end is also usually produced by dedicated speakers, as the high-frequency vibrations are more easily produced by specially designed speakers that can reproduce them better than normal 'woofer' speakers. These are usually called 'tweeters' (a specific type) or horns, which are very high specification and specially designed speakers mounted on a horn-shaped flare.

Speaker repair shop

A woofer

Speakers of all varieties are often called 'drivers' as a general term. You could ask what drivers are in a speaker cabinet and this would include any types that are in it.

A 'bullet' tweeter (left) and a horn tweeter

A horn flare. The driver screws onto the thread at the back

Cabinets (or enclosures) that house speakers come in many shapes and sizes. Often, the design is a major part of the sound that emanates from the cabinet (cab or 'box' for short). Generally, horns and tweeters don't require a special cab as they're self-contained. In fact, some venues such as cinemas often have them simply mounted on top of cabinets and not even fitted inside. Generally, woofer type speakers will point out of the cabinet so that they can be aimed at their target but sometimes they may face inward, downward or at an angle depending on the cabinet design; this is often the case with subs that may have cabs designed to produce resonance. Ports are also sometimes used, which may be in the form of a tube or slot to generate energy and extra response produced when the cone moves back into the cabinet. Often, cabinets will contain a range of drivers, even sufficient to cover all frequencies (called 'full-range cabinets'). A cabinet that has drivers covering different frequency ranges will have either connections for separate amps to drive each type (bi-amped) or a built-in electronic system called a 'crossover' that controls the frequencies sent to each driver, ensuring maximum efficiency and protecting drivers that could be damaged if they receive frequencies outside their operating range. Bi-amped or separate cabinets would use an external crossover (see Chapter 10).

Speaker cabinets come in various shapes and sizes

Speaker systems vary widely in their design. For example, you could have traditional 'stacks', which are speaker cabs arranged in stacks. Systems may be flown, which involves hanging some or all of the cabinets from suitable suspension or 'flying points' either in a venue or from a specially erected tower on outdoor gigs, and often a combination of both. Recent years have seen the popularity of 'line-array', which usually consists of sub cabs along with cabs that are very compact (in fact, they sometimes don't look big enough to do the job) but very efficient and cover most of the frequency range. This system is very flexible as it can be quickly rigged in various formats for a variety of venues on a tour and generally takes up very little space compared to older stacks. Installed systems often have to fit in with architectural considerations and so could have speakers anywhere, often in less than suitable locations. Some speaker companies now provide software to help you design and aim your PA rigs to suit venues, so it's worth checking out a few web sites to see what's available.

The power of speakers and amplifiers is generally rated in watts. Sometimes this can be misleading as there are various ways of quoting wattage. Generally, we need to use watts RMS (although I know that some scientists aren't happy about this, it is a generally accepted term of reference). In the professional world it is the term of reference used and it is understood that the average power of a 1000W amplifier is 1000W. Sometimes you may see other measurements used such as 'music power' or PMPO (Peak Music Power). This is the highest that the output will go before damage and can vary between appliances and manufacturers even though the same RMS rating applies. As such this shouldn't be used as a guide. Average power is what you'll be using, so you should be well within operational limits when peaks do come along.

It's quite possible, and happens a lot, that speaker and amplifier ratings are misunderstood and so they are mismatched, which can result in poor performance and damage. The best advice is to take recommendations from manufacturers but it isn't a good idea to assume that (as often happens) you need speakers that can take more power than the amp is giving out (e.g. a 300W speaker and a 200W amplifier). This can be very dangerous as you could easily drive the amp so hard that it distorts (clipping), which may then damage the speaker due to a distorted signal that the speaker cannot reproduce. The design of a system needs to take a number of things into account.

Speaker cabs flown (a) and stacked (b), line array speaker hang (c), and bass cabinets ground stacked (d)

A trashed woofer

It's always good to have plenty of power to spare for all parts of the system, which we often call 'headroom'.

Another important measurement for amps and drivers that you really need to know about is 'impedance'. Simply stated, this is the resistance that the driver puts up or the pressure that the amp can put out. A rating is usually stated for amplifiers and speakers on a label attached to them. If more than one driver is fitted into a cab then the cab rating may be different from the drivers and the cab will usually have a rating as well as the drivers. It's important to know because a mismatch will result in the termination of your gear. We measure impedance in ohms, again after a clever chap Mr Greg Ohm, a Bavarian mathematician and physicist, 1789–1854. Ohms are also used to measure resistance in electrical components and use a symbol.

The Ohm symbol Ω

To use the water analogy again we might see the amp as a pump that pushes water down a pipe (the wire) and the speaker could be a sprinkler. The pump (amp) can put out a fixed volume of water at a fixed pressure and provided the sprinkler (speaker) is working at its best at this level everything is fine with the world. If, on the other hand, the sprinkler needs more water to operate (it's offering less resistance) then the pump has to work harder to keep up the same flow. If it can't keep up, the sprinkler will fail.

If the sprinkler offers too much resistance then the amp isn't working at full efficiency as it's holding back some of the flow it has available.

Back in the real world this translates that you should never connect a speaker (or cabinet) to an amp if the speaker has a lower impedance than the amp is rated at; if the amp is rated as running at a minimum of 8 ohms then you shouldn't use a 4 ohm speaker with it or you'll blow it up!

You can work the other way round but you'll get less power out of the amp. An amp that will run down to 4 ohms will produce less power in watts when connected to an 8 ohm speaker than a 4 ohm speaker as the speaker will impede the flow to some extent.

The way that drivers are wired in a cabinet will have an effect on their impedance. Using a pair of 8 ohm speakers wired in parallel will result in an impedance of 4 ohms, which is half of the original.

If the same speakers are wired in series then the impedance is doubled to 16 ohms.

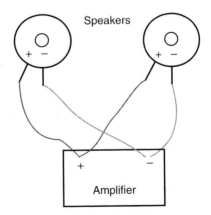

Parallel wired speakers

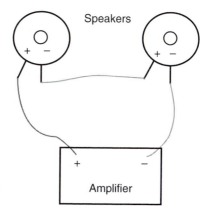

Series wired speakers

You can also wire in series–parallel with more speak-ers and keep the same impedance for the cab as for the speakers (useful in a 4 × 12 guitar cabinet).

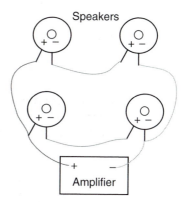

Series/parallel wired speakers

Generally speaking (no pun intended), you shouldn't wire speakers differently unless you know exactly what you're doing; calculate all of the impedances and make sure that they work before you start changing things around. Better to stick within the recommended guide-lines provided by the manufacturer.

When looking at amplifier and driver specifications you need to be sure both the impedance and wattage ratings are compatible as an amp that puts out 500W at 8 ohms is not at all the same as 500W at 4 ohms. I have a power amp that is rated:

500W RMS per channel into 4 ohms
300W RMS per channel into 8 ohms

This illustrates the difference in power output versus impedance.

The 'per channel' means that each output channel is rated as such. This particular amp is stereo and so it has two channels.

There is another way of running amps and speakers called 'bridge mono', which basically connects both outputs of a stereo amp and uses their combined power to send a single output, which in the case of my amp above is rated at:

1000W RMS into 8 ohms (bridge mode)

If you want to run an amp in bridge mode then you should only use one that has the facility to do this. Use its onboard switch (if provided) and follow the manufacturers' instructions to the letter to avoid destroying any gear. There is an obvious increase in power although not at the lowest impedance that the amp is capable of. The basic idea is that the channels (really separate amps inside the casing) are connected together and you then take the positive terminals as the output to connect your speakers to. If this sounds a little strange it is, as the amps are connected out of phase with each other, so

The bridge switch on a power amplifier

the polarity is different. Implementation of this can also vary between manufacturers, so I wouldn't advise doing it unless you have the manual or instructions from the manufacturer.

Bridge (or bridged) mono is most often used to drive the low end or sub-parts of a PA as they require the most power to drive them. Bass frequencies are essentially monophonic in content, so a stereo amplifier is not necessary.

At this point the quick thinkers among you will have figured that there are ways of wiring speakers and running amps in bridge mono that can provide a lot more power than running them in stereo, this is very true but if you accidentally switch into stereo or connect anything up incorrectly you'll be frying tonight, so be warned!

Drivers can take a lot of wear and tear. They often wear out, burn out or seize up and this isn't always the end of the road for them. It's possible to repair speakers and there are many specialists who will do this for you. Horn drivers consist of replaceable components such as diaphragms and voice coils as do some tweeters (although some are sealed and disposable) while woofer type drivers can have the chassis salvaged and 'reconed', which fits all new components in the still-good chassis. It is possible to purchase kits to do this yourself. There are also kits to replace various parts such as the flexible surround that supports the cone.

Speakers are often referred to by their diameters. A cab with a pair of 12-inch speakers would be a 2 by 12 (2 × 12) or this may be shortened further to a 212 (two-twelve). Even though metric measurements are common the old inches are still used where speakers are concerned. Popular woofer sizes are 12″, 15″, 18″ and sometimes as big as 21″ or smaller sizes such as 8″ or even 4″ are found. The drivers on horns (compression drivers) are also often referred to in inches, the most common being 1″ and 2″.

Tweeters are very small speakers designed to reproduce high frequencies. There are also Piezo tweeters that use a piezo crystal to provide the sound and ribbon tweeters that use a foil ribbon to produce the sound.

Damaged speakers can be repaired

Inside a compression driver

Some speakers use waveguides (such as molded horns) to focus the sound more consistently.

Drivers often have soldered connections but some may have push-on connectors. I have known these to work loose and cause consternation and even the dismantling of amplifiers when searching for the fault. Whenever you change one, make sure that the connectors are securely fitted.

Connections

The cables used from amplifiers to speakers are usually heavy-duty copper cables that don't require screening in the same way as mic or instrument cables do as they aren't as susceptible to interference. There are 'proper' commercial speaker cables but decent two-core electrical cable can also be used.

Cables from mixer/processors to amps should be good-quality screened cables as they are susceptible to interference.

Power to the people

You may use powered speakers that have a properly matched amplifier attached to the cabinet and often an internal crossover system to protect and ensure efficiency from the drivers. These are absolutely fine except that if an amplifier or speaker breaks then the whole unit is out of action until repaired. They are, however, designed to work together, so you don't have the problems of matching amps and speakers.

Powered speakers. Note the controls and cooling fins on the back of the built-in amplifier

A small powered mixer

Small systems may also use a powered mixer or mixer amp that provides mixing and power amplifier systems in one box. These systems save carrying separate amps but usually only have stereo amps built in, fine for small gigs. The only real drawback here is that you have to run speaker leads from the mix position and that could mean long runs of heavy cable.

8

Microphones and DI boxes

In order to amplify the sound and reproduce it in the PA system we first need to obtain a real-time impression or copy of it that we can use. We usually do this in one of two ways. One method is to use a microphone that absorbs the vibrations and converts them into electrical impulses. Another way is to intercept the instrument's electrical output signal and use that, with a piece of equipment called a DI box. It's useful to have a good idea about the various types of microphones and DI boxes available and what their different capabilities are.

The proximity effect and axis response

Some microphones exhibit what is known as the 'proximity effect', which can work for or against you depending on if it's being used correctly. In essence what it does is to increase certain frequencies (usually lower ones) when placed closer to the sound source. This is most easily demonstrated by singers who get very close to their microphone on some, or indeed all, of their vocal performance.

Axis response is similar and depends on whether the mic is placed pointing directly at the sound source or at an angle (off-axis). You are more likely to see a reduction in some frequencies when used off-axis.

Some vocalists have very good mic technique while others are pretty bad and cause feedback by cupping the mic or holding it incorrectly. There's not a lot you can do about this other than explain that they will get a better sound if they hold the barrel of the mic rather than the live end. I've also seen vocalists who will hold the mic too far away (maybe through a lack of confidence in their voice), which means that the vocalist won't hear their monitor very well and their channel will need turning up at FOH. If they then decide to sing close-up it will suddenly be too high. These are problems that can be cured with better mic technique but if you're running the mixer then you'll need to watch for them and react accordingly.

Microphones

Most people are familiar with microphones (mics for short) as they are pretty visible when used by vocalists in particular. But there are different types of mics and variations of the types that have special uses such as wireless radio mics. The most common connector for professional microphones is the XLR. Occasionally, another type of connector may be used but if so there will generally be an adapter provided to connect to an XLR plug.

We're going to look at the types that you're most likely to come across on live shows. The most common is the dynamic, so we'll look at it first.

Dynamic microphones

These are the type that most people will be familiar with as they are seen almost daily on TV when vocalists sing into them. They are available as wired (with a wire connecting them to the mixer) or wireless (using radio signals to transmit to a receiver that is plugged into the mixer). If you see someone using a radio mic it may have a little wire hanging out of the end, which is its antenna. They are usually a bit bigger than wired mics as they have to carry circuitry and a battery to power it. This also means they often need larger mic clips on their stands.

Dynamic mics tend to be of rugged construction, with their internal components suspended in antishock mountings. They sound good, are fairly cheap and are easy to use. The basic principle of the dynamic is that sound vibrates a diaphragm inside the mic, which is attached to a coil. The coil moves with the vibration of the diaphragm within a magnet's field. This produces a voltage in direct

A variety of dynamic microphones

A dismantled dynamic mic showing the body, shield and diaphragm capsule

relation to the sound pressure, which is then sent to its output.

Dynamics are used extensively for vocals, drums, and 'micing up' amplifiers such as guitar amps but they can be used for almost any application. The most popular polar pattern for dynamics is cardioid, which provides good focus and helps to reject feedback.

Condenser microphones

Widely used in recording studios the condenser micro-phone has a better response to a wider range of fre-quencies and reproduces sound more faithfully than the dynamic. It is generally not as tough as a dynamic and also requires power in the form of either 'phantom power' supplied by the mixer or battery power (often by a bat-tery fitted into the mic body). Phantom power is usu-ally the best. They work by using a very fine diaphragm usually coated with a thin layer of gold, which works in conjunction with a fixed backplate, separated by a small air gap. A voltage is applied to the diaphragm assem-bly, turning it into a capacitor (condenser). When the diaphragm is vibrated by sound waves the capacitance changes as the distance between it and the backplate changes. This change is then used to provide the out-going signal. As the diaphragms don't have to support a coil assembly they can operate very quickly and accu-rately.

There is another variety of capacitor microphone called an 'electret' mic that works in a slightly different way. It uses a material that has a permanent electrical charge and so doesn't require a polarizing voltage. It does, how-ever, normally carry a pre-amplifier, which does need power. They are often very small and can be used as lapel type microphones but do not generally perform as well

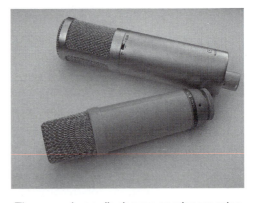

These are large diaphragm condenser mics

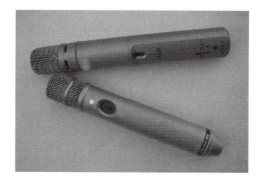

These are small diaphragm condenser mics

These are very small electret lapel (or lavalier) type mics with their power supplies

as high-end condensers. They can, however, be used as wireless mics due to their small construction.

Condensers come in a variety of shapes and sizes but are not generally available as wireless mics due to their power requirements. They are often used to capture high-frequency content such as cymbals and hi-fidelity acoustic sounds such as piano, violin or choral sections. Because of their high-quality reproduction they can be a little awkward to use close to PA speakers as they are more prone to feedback if not set up correctly. If too close to high-pressure sounds such as inside a kick drum they can also overload and make nasty noises. In a rock band setup though they will usually be found on hi-hats and drum overheads.

Some condensers have switches fitted such as 'pads' to attenuate (reduce) the signal and high-pass filters to reduce frequencies below a certain level (such as 70 Hz) to remove unwanted rumble. Some also have switchable polar patterns so that the mic can be used as cardioid, omni or figure of eight for instance. Condensers are generally available in most polar patterns to suit their intended application.

Boundary (pressure zone) microphones

These mics are different in appearance to most others as they are usually a plate, often with a mesh screen over the top, which can be placed onto flat surfaces such as floors or walls. They work by picking up the sound reflections from the plate using a small microphone that is set a short distance off the plate. They require power in the form of phantom power from the mixer or have the facility for a battery, often in a case attached to the cable.

There are two types of boundary microphone. The Pressure Zone Microphone (PZM) and Phase Coherent Cardioid (PCC) differ slightly in their design. The PZM has an omnidirectional capsule above the boundary and picks up sound in a hemispherical pattern above the surface. The PCC has a supercardioid pattern and rejects sounds coming from the sides or rear and so is more directional.

Boundary mics are often used to pick up areas of sound such as a group of singers, actors or musicians, but they are also used for specific jobs such as inside kick drums or pianos. I recently used one placed on a stage floor so that the taps of the dancer's shoes could be heard at a festival of Irish music.

A boundary type mic commonly used for kick drums

Microphones are generally supplied with charts of their frequency response curves, usually available from the manufacturers websites along with other specifications such as the level of sound that the mic can withstand before distorting (Max. Sound Pressure Level or SPL).

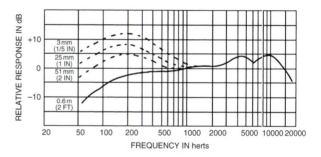

A typical microphone frequency response chart

Polar patterns

Microphones pick up sound from an area around their capsule (the business end, usually protected by a screen), but this area is usually tailored by the manufacturer to suit the application intended for the microphone to be used in. There are a range of generally used shapes that give a good idea of where the mic will respond well to sound and also where it will reject it; this is important when considering mic placement as the rejection areas can be used to help prevent spillage from other sound sources and also cut down on the risk of feedback from monitors and PA speakers.

Here is a list of the most popular patterns you will come across:

Cardioid: The name means 'heart-shaped'. There are variations of this shape such as supercardioid and hypercardioid. The cardioid pattern is probably the most common in use as it provides good reception from the front of the mic and good rejection from the back. It will however take in some sound from the sides.

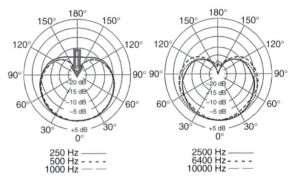

Typical cardioid patterns (as viewed from above, arrow shows position of mic)

Hypercardioid: A tighter version of the cardioid pattern. It has better rejection of sound coming from the side due to its better focus. It may, however, pick up a little sound from the rear, so care should be taken if using with monitors.

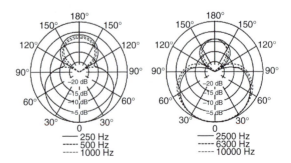

Typical hypercardioid patterns

Omnidirectional: Will pick up sound from all around its diaphragm. This is a pattern usually available with some large diaphragm condensers (patterns such as cardioid are also referred to as 'unidirectional' as they receive sound from only one direction). This pattern is more suited to recording than live sound due to its sensitivity in all directions but it may be used for large ensembles such as choir or orchestral.

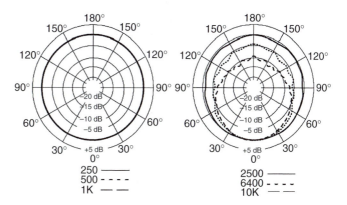

Typical omni pattern

Bidirectional: Also often called 'figure of eight' this pattern will accept sound from two areas usually opposite sides of the mic, again this is often a pattern available on some condensers. It will pick up sound from opposite sides of the mic and reject it from the other sides. Not generally a live mic.

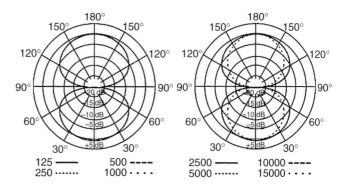

Typical bidirectional (previous figure) pattern

Feedback

The bane of a sound engineer's life (and anyone else who is in the room when it goes off big time), feedback (sometimes called 'howlround') is not a pleasant sound unless created intentionally by a guitarist for effect. It occurs when the sound source is amplified and then sent out through a speaker, the sound coming from the speaker is then received again by the microphone (or pickup) and amplified again in a continuous loop. This is the squealing that you sometimes hear at live events. Microphone pattern and placement is important in helping eliminate feedback as you can help to ensure that they aren't likely to pick up the sound from speakers that they'll be feeding into. Probably the worst offender is volume and proximity to speakers but every situation is different. The more mics and speakers you have, the more likely it is that you'll generate some feedback. If the PA speakers are well positioned and controlled then it's likely that most feedback will be coming from onstage monitors. In this case you need to look at this system in detail and possibly use a bit of frequency cutting with a graphic equalizer to sort things out (see Chapter 9 for more information).

Microphones properly placed to help prevent feedback from the monitors

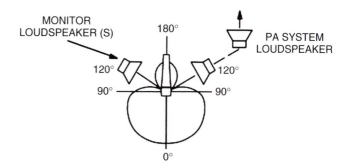

MONITOR
LOUDSPEAKER (S)

180°

PA SYSTEM
LOUDSPEAKER

120°

120°

90°

90°

0°

Here's a diagram showing how to position microphones, monitors and PA speakers to avoid the mic picking up the amplified sound and thus cutting down the likelihood of feedback

In general, the rejection area of the microphone (usually the rear) should be facing any monitor or speaker. If the microphone's responsive area is pointing that way it may pick up some of the amplified sound and feed it back into the system causing a feedback loop.

DI boxes

DI (pronounced dee-eye) stands for direct injection and this is what DI boxes enable us to do, inject a signal directly into the system, more specifically to conform with the system's parameters. Most often it is used to take an unbalanced and/or high-impedance signal from an instrument and feed it into the system as a balanced, low-impedance signal on an XLR connector, the same as the microphones. Many instruments such as synthesizers, bass guitars and acoustic guitars with built-in pickups are often best when used with a DI rather than being mic'ed up. The signal will be what is coming directly from the instrument, uncolored by any amplifier or processor that comes later on the stage in the musician's backline setup.

A DI box is connected to the output (or outputs if stereo) of the instrument using a 1/4″ jack connector and then another jack lead is used to link from the box to the backline amplifier. An XLR lead is taken from the box to the stage end of the snake, which is then sent to the mixer. Facilities often exist on the box to attenuate (reduce) the signal as some instruments may send a signal that is too 'hot'. These usually take the form of 'pad' switches that reduce the signal by a fixed amount when depressed. As an example you may find two pads rated at 20 dB each. Depressing one will reduce the signal by 20 dB but adding the other will reduce the overall signal by 40 dB. It is also usual for DI boxes to have an 'earth, or ground-lift' switch that can be used to take the instrument's own earth out of the path if there is any hum created due to the instrument being connected to the mixer and backline amp both of which may have their own grounding points. It is normally left in, that is the ground isn't lifted unless necessary.

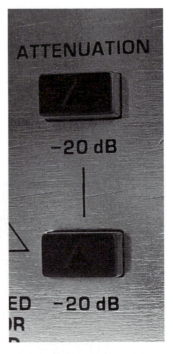

DI pad (attenuation) switches

Most DI boxes are single channel and designed to be used with one instrument usually on the floor of the stage or on top of a backline amp. There are also stereo and multiple-channel boxes available, the larger ones usually residing in a rack.

DI earth (ground) lift switch

DI boxes come in two types, active and passive.

Passive DI boxes

Passive DIs perform all of the necessary functions required in a small, lightweight package, which is usually priced very reasonably. They don't require any power at all and as such can be used with a mixer that doesn't have phantom power facilities. Passive DIs can be found in multiple-channel configurations such as on subsnakes for keyboards.

Passive DI boxes being used to connect a drum machine into the PA system

Active DI boxes

Active DIs require power either in the form of phantom power from the mixer or from an internal battery. Because of the powered system they are capable of producing a wider frequency response by controlling the input versus output impedances, thus preserving better signal integrity for the connected instrument. Because of the extra circuitry involved, active boxes are often a little larger than their passive counterparts but there are also some very small active boxes around. Multiple-channel rack mounted boxes are often active and powered by mains electricity.

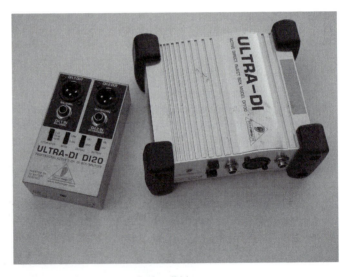

Active DI boxes

DI boxes can occasionally solve tricky problems even where a DI box wouldn't normally be used such as using the box as a jack to an XLR adapter. A trick I often use when earth noise is present in an amplifier's XLR output that is going straight to the mixer is to put a DI box inline (it must be one with XLR in and out) then use the earth-lift switch. You could also use the pad switches to reduce a hot signal.

A general idea of what to use where

Although everyone has their own idea of what to use and where, you probably would like to have some idea where to start; here's a list of possibilities. I was once told by a top engineer/producer that if he had a boxful of decent dynamics he could mic up anything. Never get frustrated if you don't have the mics you want, just use the ones you have wisely.

Note that these are just general guidelines!

Drums
Kick: Dynamic or boundary mic
Snare: Dynamic mic
Hi Hat: Condenser mic (pencil type)
Toms: Dynamic mic
Overheads: Condenser mics
Percussion: Condenser mics
Bass guitar: DI or condenser
Keyboards: DI boxes
Electric guitar: Dynamic mic on cab
Pedal steel: DI box or dynamic mic on cab
Vocals: Dynamic mic
Acoustic guitar: DI box or condenser mic
Acoustic piano: Condenser or boundary mic (may need more than one)
Strings: DI box if pickup on instrument or condenser mic
Woodwinds: Dynamic or condenser/boundary mics for sections
Brass: dynamic or condenser/boundary mics for sections
Choral: Condenser or boundary mics

Micing techniques

There are myriad ways of putting microphones on a stage to amplify artists but certain techniques have become a standard as they are tried and tested and work! You don't have to use these methods and often the availability of microphones and the area you have to work in will dictate your approach, but here are some useful techniques:

Close micing: This is where a microphone is fixed in a position very close to the sound source such as in front of a speaker in a guitar amplifier/cabinet or close to a snare drum.

The advantages are that you get an extremely good focus on the sound and the level will be very good provided that the source is at a good level.

You don't however get any ambience from the room. There is a danger that the mic may get damaged by a stray drumstick hit or may not be able to handle the sound level (Max. SPL).

Close micing a snare drum

Ambient micing: Good where you have a group of performers such as a choir or string section and want to get a coherent sound with all performers present in the signal. This technique involves micing from a distance to capture all performers and/or ambient sounds. Also used for drum overheads to pick up the whole kit especially the cymbals.

Requires a good quality sensitive mic usually a condenser (or several) works best or boundary mics for some areas if there are suitable mounting points.

Placement is very important and it may be difficult to achieve this easily. You also have to be aware of potential feedback problems as the microphone will probably need a high level of amplification as it is a distance from the sound source, you may also need several mics to cover the whole area required.

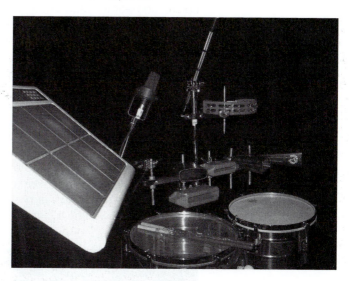

This percussion rig uses ambient mics to capture groups of instruments

Stereo micing: Can be done close up or at a distance requiring the use of two mics, one for left and one for right. Ambient sources benefit from stereo micing if large areas are covered, enabling the actual positions to be represented correctly when amplified through the PA system by using the mixer 'pan' controls. Close miced instruments such as drums also benefit from some stereo positioning (panning) in the mix, particularly toms and overheads. There are various ways of setting up ambient stereo mics, the most common for live work being a simple A–B pair that are separated width-ways and aimed at the sound source.

Another often used stereo arrangement is the X–Y pair. This consists of a pair that are crossed at right angles, one on top of the other, to help eliminate phase problems.

Double-micing: Sometimes instruments are double-miced for specific reasons, the most common being to add certain properties such as various tones or sounds available from different parts of the same sound source. As an example a snare drum sounds different if miced from the top than from the bottom as the wire snare is underneath. Some engineers will place a mic above and another one below (sometimes different types to capture different frequencies). If they do this then it creates a problem as the top mic senses the sound wave opposite to the bottom one (imagine how the drum skin, when hit, moves away from the top mic and toward the bottom one at the same time). This makes them 'out of phase', which is corrected using a switch on the mixer, or the connectors on the mics may be wired in opposite polarity to each other. A kick drum could have a mic inside and another outside to pick up the beater sound as well as the drum sound. A grand piano may well have mics on the higher and lower strings inside its lid. Any double-micing may cause phase problems, so you should always check; if the sound is down in level it could well be a phase problem.

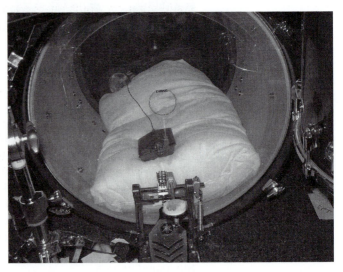

This kick drum is double-mic'd with a boundary mic inside and a dynamic at the cutout in the front skin

Microphone and DI: You may sometimes use both a mic and a DI on an instrument (such as acoustic guitar) and mix the two sounds together to produce a more pleasing tone. This can cause phase cancellation problems, so be careful.

Whatever microphones you use and whatever micing techniques you decide on, the main objective is to make the artist heard as clearly and accurately as possible. Always bear this in mind as a priority over everything else.

Microphone cables

The cables we use for microphones and DI boxes are 'balanced' using two internal cores and a screen. They will have an XLR plug on one end and an XLR socket at the other.

Balanced wiring helps to eliminate any external interference and has better signal integrity, particularly over long runs, when compared to unbalanced lines. As it uses two cores plus a screen the signal is carried along the cores and doesn't rely on the screen to pass its signal. If a 'ground loop' problem is causing a hum in a balanced line it's possible to remove the screen at one end in order to break the loop. If the line has phantom power running through it you can't do this though.

Instrument cables

The cables used for instruments are generally unbalanced and only have a single core as they use the screen to carry signal as well as the core. They work well over small distances but are more susceptible to interference than balanced lines, particularly if worn or damaged.

9

Monitoring

Monitoring, often called 'foldback', is what we use to enable the artists to hear what they need to when on the stage so that they can perform correctly and in time with other artists or any sound or music being played such as backing tracks. The PA system may be a fair distance away from them and what they hear from this will be mainly reflections from the opposite wall, which will have a delay due to the sound traveling there and back. This is not a good source of reference, so artists have a more direct way of hearing the sound using speakers on the stage facing them or in-ear monitors similar to the type used with personal stereos. Vocalists particularly need a good level of their own voice to be audible to them as it can be very difficult to pitch correctly if you can't hear your own voice, such as in front of a loud band.

The drummer should hear this! Making sure the drum monitors don't fall over by strapping them together

Some instruments such as keyboards may not use a backline amplifier so their players will rely entirely on a good monitor mix to hear them at all. The bigger a stage is, the further away the artists may be from each other, so the timing and mutual feel of the music may suffer simply because of the short time it takes for sound to travel across a stage. If important sounds are sent instantly to each musician then they will be able to retain a good sense of timing and feel as if they were positioned close together.

Before the sound sources onstage are sent to the mixer out front they are split into two outputs, one of which goes to the monitor mixer and the other goes down the snake to the FOH mixer. If recording or outside broadcast (OB) requires yet another signal feed then another split can be made here or at the FOH mixer. Often mixers will have 'direct out' sockets on every input channel where these signals can be taken from.

There may be a lot of wiring in a monitor system

Problems

If you've done any work at all with microphones and amplified speakers then you'll probably have discovered feedback when the mic picks up the sound from the speaker and sends it back into the system where it is amplified again until it rapidly causes some horrendous squealing noises. Monitoring is a system that almost invites this behavior by putting lots of speakers on the stage pointing at lots of microphones and sound sources, which is a really good recipe for creating feedback. A lot of this can be avoided if the stage acoustics don't reflect the sound back toward the monitors (a general term for monitor speakers) too much. Careful positioning of the monitors themselves also helps. If a monitor is pointed toward the performer from behind his/her microphone stand it will (in most cases) be in the area where the mic doesn't pick up sound due to its polar pattern (see Chapter 8 for more information). Positioning and stage acoustics won't prevent all risk of feedback, however, so the usual method

A typical rack of 31-band graphic equalizers

of eliminating or reducing the likelihood of it is to put a graphic equalizer in the signal path. The type of graphic equalizer used most often has 31 separate bands of frequencies that allow a good degree of fine-tuning when only a narrow band of frequencies has to be reduced.

There are also units that are called 'feedback eliminators and graphics' (the abbreviation for graphic equalizers) that have indicators to show high signal levels that are likely to feed back, which they achieve by using electronic detection circuitry. The idea here is that you can see where there may be a problem and adjust settings yourself or allow an automatic device to take over from you. Many monitor engineers though still prefer to use their ears and 'manual' graphics to cut any relevant frequencies by the amount they feel is necessary. This also gives a good and instant visual indication of the overall equalization of every monitor mix.

Monitor types

Any type of speaker can, in theory, be used as a monitor speaker but we should try and use something with a suitable frequency range for what we expect to send to it. A sub-bass cabinet may not be much use for a vocalist but would be ideal for a drummer who wants to 'feel' the kick drum when he/she uses it. There isn't really a need to send stereo sound to monitors, so we can use mono speakers. The norm is to use single full-range cabinets that cover a broad range of frequencies either singly or in pairs but just using a mono feed.

Wedges

The most frequently used type of monitor cabinet is the wedge, which is so called due to its shape. They slope at an angle useful for aiming straight at an artist's head while sitting directly behind a microphone stand (the wedge that is). The design of the cabinet may also incorporate a shallower angle that is useful for aiming when placed on top of a platform (such as a flight case) so as to be nearer to the vertical. They can also be used standing on one end or even mounted on poles if they have the relevant 'top-hat' fitting.

Although they can have fitted whatever speakers are necessary they generally consist of a 12″ or 15″ woofer and a horn with either 1″ or 2″ compression driver and contain a built-in 'passive' crossover. This is a device that will split the incoming signal frequency bands and send the correct range to the speakers (this prevents potentially damaging low frequencies going into the high-frequency horn drivers). It doesn't require any power hence the term 'passive'. They will have sockets that allow the signal to be brought in and also taken back out again when two or more are used together. Some monitors have the facility to be bi-amped meaning that the different drivers can be powered by separate amplifiers and an external active crossover may be used to separate the frequencies fed into the drivers (speakers).

On the semi-pro level it's common to see self-powered wedge monitors with built-in graphics, volume controls and outputs to add another unpowered monitor that can be driven from the powered wedge. This gives some control to the artist when a full monitoring system is impractical.

Wedge monitors

Sidefills

On large stages where an artist may cover a wide area there can be gaps in the monitor coverage. Many artists like to wander along the front of the stage from one side to the other and so they can be a long way from their wedges. In this case it's usual to add extra monitoring in the form of sidefills. These are basically full-range speaker stacks placed either side of the stage near to the front facing toward each other covering the general downstage area. Sidefills can be placed anywhere that's necessary but feedback can become more of an issue as they will generally be picked up fairly easily by a handheld microphone, particularly if an artist walks close to them. The mix in them will often be different as they may need more of the backline instruments because they will sound quieter at the front of the stage and may need some bolstering.

Typical sidefill stack

Drumfills

Not a roll around the toms, a drumfill is the drummer's monitor, which is usually much bigger than a single wedge. The reason for this is that drummers often like to have their drums fed into their monitors and (especially rock drummers) like to 'feel' the kick drum. To achieve this, it's common to have a sub (bass) speaker as well as full-range speakers. This can be as big as a large stack (or two). Drummers also often need a high level of bass from the bass guitar and keyboards, so the sub will be doing its job most of the time. If a drummer uses in-ears he/she may also need a sub speaker so that he/she feels the sound pressure of the kick drum. There is an alternative device that attaches to the drummers stool and 'thumps' it when it receives the signal from the kick drum channel. There must be many other uses for them!

Typical drumfill consisting of bass cab and top cab

In-ears

If every artist used in-ear monitors then the monitor engineer's life would be so much easier and may even seem like an attractive career! In-ears are very similar to the small ear-plug type of headphones that you get with personal stereos or mp3 players but the better ones have specially made custom earpieces that are molded to fit the user's own ears. They can help to block out external sound, allowing the user to have a more comfortable monitor level that they can control for themselves using the pack that they plug into. Wired in-ears are available for artists who are in a fixed position such as drummers, while radio type can be used by more active artists. The chance of feedback in them is virtually nil as the possibility of a microphone picking up their signal is very small even if fairly close. They do need a little care as they can be quite delicate, and spare molds or earpieces should be carried at all times. You need to fit batteries regularly as failure during a show is not at all desirable; in fact, changing them before every show is usual on tours. The thing that can cause problems for some performers are the wires coming from the earpieces to the belt pack but using a little theatrical body tape enables them to be fixed securely even through costume changes. The monitor engineer will also benefit from having his/her own set in order to hear the mixes accurately.

The user will probably have to go through a period of time getting used to using them as it can be a difficult thing to do initially, especially for someone used to using wedges. Perseverance should lead to success and a better (lower) level of sound onstage is required, which is always a good thing for anyone's hearing. In order to reduce the feeling of isolation created with in-ears it's a good idea to add an audience mic so that the general ambience can be added into the user's mix. It's also worth considering the use of a stereo system that will add a further feeling of dimension.

Some performers like to have in-ears and speakers (greedy I know) so that if they feel like it they can take out one (or both) of their earpieces to hear more of the natural ambience or audience sound. If you're the monitor guy and the lead singer wants in-ears, wedges and sidefills then you'll just have to deal with it, keep smiling and look forward to the end of the gig/week/tour.

Wireless in-ear monitors showing transmitter, receivers and earpieces with their molds

Monitor mixing

The monitor mixing console has certain basic requirements that may be fulfilled by a 'standard' mixer or a special monitor mixer may be required. The basic facilities that it needs are enough input channels with proper control over gain, EQ, and so on for all of the sound sources and enough output channels for all of the required monitor mixes. It should also have the ability to control the level of each input signal that is sent to each output. An FOH mixer will usually have auxiliary (aux) controls that can send any incoming signal to an aux output that can be routed to a monitor path if required. It may have a limited number, not enough for the mixes required onstage hence a dedicated mixer would be required. FOH aux sends would also have to be what we call 'pre-fade' if they were to be used for monitoring. If they were 'post-fade' they would go up and down with any FOH mixing performed with the faders. Often, FOH mixers have at least some fixed 'post-fade' aux sends that are used for sending to effects but are not very good for monitoring from (although it is possible). A dedicated monitor mixer will have many more sends that are designed to be used specifically for monitoring and will also have facilities for the monitor engineer to listen to any of the incoming signals or outgoing mixes that are being used. This is usually performed using a 'listening wedge' or in-ear monitors, which are the same type that the artists are using.

Here is a brief example of using this technique:

The lead vocalist asks for more of his/her vocal and less kick drum in their in-ear monitors, so the engineer inserts his/her own in-ears and listens to the lead vocalist's mix (by selecting it on the mixer) and adjusts as necessary. The vocalist then gives the thumbs-up signal when it is good.

Now the drummer wants more kick drum and a little more bass guitar in the drumfill, so the engineer takes out his/her own in-ears and calls up the drummer's mix into the listening wedge, adjusts and watches for the drummer to nod when it's what they want. The engineer can hear the relative levels as he/she adjusts, giving them a good point of reference with regard to what each artist can actually hear.

The different mixes are called up by the monitor engineer using the 'solo' system on the mixer that enables any channel or output to be sent to the listening section.

Monitor world on a festival can be a busy place

Huh, yeah, tsss, mmm, aah, one, two

The monitor engineer has to bring all of the signals into the monitor mixer in the usual way and set their levels and EQ before sending them out to monitors. He/she then has to make sure that each monitor has been EQ'd so that it sounds clear and without feeding back when used. To do this they will walk out to each monitor position in turn (before the soundcheck of course) and make funny noises down the mics. By doing this they are trying to recreate some of the frequencies that may feed back and are also ensuring that the sound is nice and clear. He/she will often try and cause feedback so that they know where it is likely to occur (frequency-wise). Then he/she can go back to the graphics and reduce the level of that particular frequency in that particular monitor. This must be done for every monitor position including sidefills and drumfills. The process of causing a speaker to feed back in order to use a graphic EQ to reduce the offending frequency range is often called 'ringing out'.

There may be many monitor positions that all require setting up and 'ringing out'

If the act uses any form of playback then it should also be available at the monitor mixer to be fed to anyone who needs it. As this will sometimes be run from FOH it may be sent to the stage down one of the return channels or a tie-line if any are being used. Tie lines are simply lines in the snake that are spare and can be used for sending signals, usually either way. If the bands are to play in time then they need to hear the track. Sometimes, playback tracks, sample loops or click tracks (metronome) are triggered by band members and they will usually be fed straight to the monitor mixer if not needed out front, or into the stage box and through the split if they are required out front as well as in the monitors.

Reverb may be used in monitor channels if artists require it and dynamic processors may also be used to enhance or correct signals coming into or out of the monitor mixer. The workings of graphic equalizers, effects and dynamic processors will be covered in more depth in Chapter 10.

Monitor engineers really have to stay focused throughout the show as they often have numerous artists to keep an eye on. They may not always get clear signals from an artist as to what they require so a keen eye, patience and a thick skin are all good attributes. A good sound engineer may get a lot of praise and thanks but not so for the monitor engineer. There are a minority of artists who are

Monitor console behind scenery

notorious for their bad treatment of monitor engineers and so tact and diplomacy are always a good starting point. If you know someone is going to give you a hard time then you have to make a choice if you are willing to live with it or not. I personally can take criticism if it's valid and I'm always willing to learn and listen to anyone else's point of view but when things get abusive (which luckily I haven't really seen) maybe it's time to go and find another gig.

Monitor mixing from FOH

On gigs where a separate monitor mixer and engineer is not practical or affordable it is often possible to put together at least one or two monitor mixes from the FOH mixer, provided it is of a reasonable specification (I have often run six monitor mixes in addition to the main mix from FOH). What we need to do this is an auxiliary send control on each channel that can be sent down to the stage, into an amplifier then out to a monitor, preferably through a graphic equalizer before the amp. Multiply this requirement by the number of monitor mixes required and if you have the facilities you can do it.

The aux sends should really be pre-fade. Many consoles have switches (often working for a pair of aux sends) that allow you to switch them pre- or post-fade. If you have a decent matrix system then this could potentially be used to send to monitors as well. If you do this, any post-fade sends will increase in level if you bring any faders up (such as for a solo), which could increase the feedback risk onstage. It may also increase the monitor level so that it is simply too loud for anyone on the receiving end.

If you have the right sends then you can create monitor mixes and run a strip of tape up the mixer alongside the sends to label who each send is going to. It is more difficult to spot where any monitor feedback is coming from if you're in front of the PA but it's usually someone on a mic, or who has just approached a mic to sing or play into it. You can 'ring out' the monitors on soundcheck in the same way as a monitor engineer does but the graphics may be out front requiring a lot of running up and down the venue or may be on the stage meaning you won't be able to get to them easily during the show. Grab a crew member from the venue if you can as assistant engineer for the gig and get them to help you 'ring out' the monitors.

If you are asked to run monitors from FOH then I think you're fully entitled to ask for a pay rise but I wouldn't hold out too much hope if I were you!

Preparing the monitors onstage before the show

10

EQ, dynamics and effects

Before launching into this slightly techy section I thought I'd point out that I'll cover the basics of what are generally deemed to be essential dynamics processes. We'll also look at their associated equipment on large shows but please don't feel that you must have all of the gadgets in order to put on a successful show. You may not really need any compressors or gates (for instance) and adding them for the sake of it could be a mistake, using them badly could seriously impact your gig.

We also apply the old rule of 'if it ain't broke, don't fix it', basically meaning that if a sound is fine already then you don't need to do anything other than set its level. Other wise sayings such as the acronym KISS (Keep It Simple, Stupid) apply, but a little more specific is the 'less is more' approach; a subtle touch will be better than a heavy-handed attitude, so use a degree of sensitivity when applying any processing to your sounds.

Equalization

EQ is the commonly used abbreviation for the word 'equalization' and applies to the term itself and the equipment required to adjust it. It's also used as a verb in comments such as 'I'll just EQ that wedge then I'll be ready for a coffee'. EQ is a very important part of what we do to get the sound how we want it on a show as it can be used for both correction and creative purposes. A quick example of correction would be to reduce a frequency range that was feeding back, while creatively it could be used to add a little high-frequency boost on a vocal to make it sweeter.

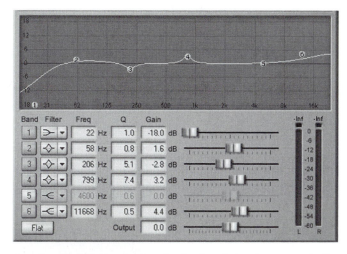

A software EQ graph. Amplitude (level) is shown vertically, frequency is shown horizontally. Each vertical graph line shows an octave spacing. The peak at 4 shows an increase of 3.2 dB at 799 Hz with its slopes (width) covering approximately a one-octave range

The audio signal will undergo any dynamic processing before it reaches the power amplification stage. Any single signal, group of signals or the whole mix can be treated with equalization, if facilities exist, at any point between the actual generation of the signal and its amplification into the PA speakers. In fact, many audio sources will undergo a lot of EQ stages before they reach the PA. If we take an electric guitar then the pickup on the guitar usually has a tone control, next the guitar effects may have EQ of their own to adjust the signal before it leaves them. Then it will go to the guitar amplifier and be EQ'd again before being picked up by a microphone, which will be sent to the FOH mixer where it could have EQ applied so that it sounds good before it leaves the mixer. It may even go through more EQ stages in the mixer if it's part of a group that has EQ applied or if it has effects added at the mixing stage. The whole mix will also have an overall EQ before it hits the power amps and anyone on stage who has the guitar in their monitor may have a differently EQ'd signal still! So you can see that it's an important thing to understand.

EQ may start at the instrument

A relatively easy to understand version of EQ is avail-able to most people in the form of a few controls on most home sound equipment and that is the controls called bass, treble and sometimes a middle, or mid-control. They are a basic form of equalization allowing you to perform broadband frequency changes to the audio over three sep-arate bands. The designer has chosen which range of fre-quencies they will control and to what extent they will change the level when adjusted. The addition of more controls that can adjust more frequency ranges will obvi-ously provide a higher level of control. If you added con-trols for lower-middle and upper-middle as well as the existing bass, middle and treble then you would have a 5-band EQ system starting with bass then moving up the frequency spectrum to lower-middle, then middle, then upper-middle and lastly treble. As you add more controls you can begin to identify their bands by naming the fre-quency around which they are centered such as 4 kHz. We know that human hearing doesn't go out of the 20 Hz to

20 kHz range, so we can keep within these limits, which makes designing an EQ device somewhat easier.

Basic EQ strip on a mixer channel

There are generally two types of equalization gadgets used in professional sound. The most popular for PA and monitor control that resides in the control racks is the graphic equalizer. The other type is the parametric equal-izer. A good example of this is that it's the type found in the vertical channel strips of mixers, but they both work in a slightly different way. You may also see parametric equalizers in racks but they are more commonly used as creative tools rather than corrective tools, which are likely to be the domain of the 'graphic'. This isn't to say that they can't or won't be used in either role, just that it is their more common role. Here is an outline of both types and how they work.

Graphic equalizers

The graphic equalizer

You may be familiar with the graphic equalizer (or simply 'graphic' for short), which can sometimes be seen in consumer apparatus such as hi-fi gear, but those are generally pretty basic in this format. There is a common requirement for their specifications when used in professional live sound that is an accepted standard and is expected anywhere. The standard is for the graphic equalizer to be a 31-band, two-channel (or stereo), 1/3 octave equalizer. They are two channels of exactly the same controls in a box, as PA systems are usually stereo and so have two sides that may need EQ'ing differently. The controls are usually arranged horizontally one set above the other so that it is easy to adjust them both together if necessary. They can be used for entirely separate channels such as when used for monitors so they provide a handy package either way. Graphics are available in single-channel format and with fewer bands but we're looking at the standard setup here.

The 31 bands are controlled by small vertical sliders that are spaced evenly along the unit and have their operating frequency points set at 1/3 octave spacing within the audible range. This provides a very good level of accuracy for adjustment with musical audio. The spacing most commonly used is as follows, starting with the lowest frequency: 20 Hz, 25 Hz, 31.5 Hz, 40 Hz, 50 Hz, 63 Hz, 80 Hz, 100 Hz, 125 Hz, 160 Hz, 200 Hz, 250 Hz, 315 Hz, 400 Hz, 500 Hz, 630 Hz, 800 Hz, 1 kHz, 1.25 kHz, 1.6 kHz, 2 kHz, 2.5 kHz, 3.15 kHz, 4 kHz, 5 kHz, 6.3 kHz, 8 kHz, 10 kHz, 12.5 kHz, 16 kHz and 20 kHz. These frequencies are what we call ISO frequencies as they are recognized by the International Organization for Standardization. There are graphics with fewer sliders covering fewer bands such as 10 bands at full octave spacing, but these obviously don't provide the same level of accuracy. By octave we mean the spacing (interval) between one musical note and another that has either half or double its own frequency. A note at a frequency of 440 Hz (A4) would have octaves at 220 Hz (A3 one octave below) and 880 Hz (A5 one octave above).

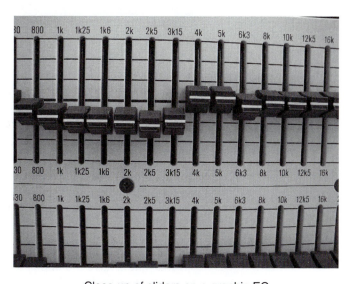

Close-up of sliders on a graphic EQ

When a slider is pushed up, the output level of that frequency is increased (boost) and when a slider is pulled down, the level is decreased (cut). Graphics often have a range switch to set the amount of boost or cut at 6 dB or 12 dB providing different resolutions with the same set of sliders. The center position of each slider is 0 and has a detent so that it can be easily located. At this point there is no boost or cut applied to the signal.

Graphic equalizers often have other useful facilities built in such as high or low filters of which two types are common. A 'high- or low-cut' filter will do what you imagine and cut high or low frequencies. It has a point at which it comes into play such as 60 Hz for a low-cut filter and then anything beyond that range is cut. A low-cut filter at 60 Hz would cut frequencies below 60 Hz and a high-cut filter at 22 kHz would cut frequencies above that.

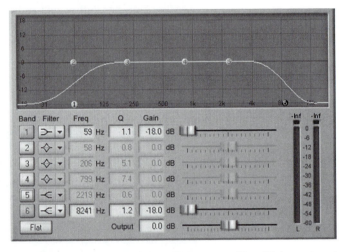

Low- and high-cut filter slopes. The low-cut filter (1) is centered around 59 Hz but actually slopes from around 150 Hz. The high-cut filter (6) is centered around 8241 Hz but slopes from around 4000 Hz

'Pass' filters have the opposite effect and will allow frequencies to pass through. A 10-kHz low-pass filter would let frequencies below 10 kHz pass through while a 60-Hz high-pass filter would let frequencies above 60 Hz pass through. The frequency at which they begin to operate may be fixed or adjustable by the user. This is what we call the cutoff or 'corner' frequency as it is the point at which the filter's shape changes and appears as a corner. Beyond this corner the filter usually slopes so that the audible change is smooth. This slope is measured in decibels per octave so a 24 dB per octave slope would be much steeper than a 12 dB per octave slope.

Band-pass filters allow a certain band of frequencies to pass through reducing or excluding those at either side.

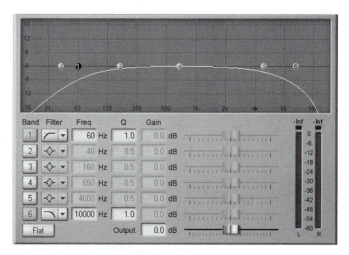

Low- and high-pass filters. The high-pass filter (1) allows frequencies above 60 Hz to pass through. Filter (6) is a low-pass filter allowing frequencies below 10 000 Hz to pass through. The bands allowed to pass are between these figures. A band-pass filter would be similar but may cover a narrower range of frequencies

The parametric equalizer

The EQ system used on most mixers is of the parametric type. There are also rack-mounted parametric EQs available and special types such as parametric valve equalizers, which are used to add a certain tonal quality to signals. Parametrics don't have the same fixed band structure as graphics but they generally use pass or cut filters and have 'sweepable' filters for the fine-tuning of some frequency ranges, generally the mid-range where important adjustments often need to be made for vocals and many instruments. They may have cut or pass filters on the very low and high frequencies with sweepable 'mids'.

The sweep system works using several controls, usually two controls plus one that can be used to adjust the shape or width of the filter (bandwidth, also often called 'Q') to cover a broad range of frequencies, or a very tight 'notch' filter. Due to its shape these are referred to as 'bell-shaped filters'.

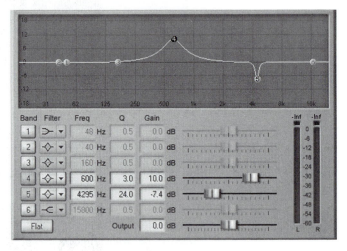

A 'bell' or peak filter (4) showing a 10-dB level increase at 600 Hz. The 'Q' factor sets the width (and accuracy) of the bell. Filter (5) has a much higher Q setting, making it very narrow (24 as opposed to 3).

The frequency control is used to sweep between a predetermined range of frequencies in conjunction with the level control. If you increase the level a little then sweep the frequencies you'll hear the EQ change as you sweep and each frequency band is boosted. Finding frequencies is often done this way, even when searching for a frequency that needs reducing as it is easier to hear when the offending one is reached. Then a reduction of the level control will accurately remove it.

Parametric EQ controls on a mixing console

Due to the adjustable bandwidth, available parametric equalizers are more 'musical' sounding, hence their use in mixing consoles and similar outboard gear.

The spectrum analyzer

If you want to 'see' what your sound is doing then you could use a spectrum analyzer, which shows an on-screen representation very similar to a graphic equalizer, that is updated constantly in order that the signal strength of each frequency band is visible in real time. This can be very useful in pinpointing any rogue frequencies that are too high or too low. A good example of this is to aid feedback identification quickly and accurately. Spectrum analyzers can often be found in specific pieces of equipment such as speaker management systems but you can also get stand-alone versions and software that can be run on a laptop or handheld computer.

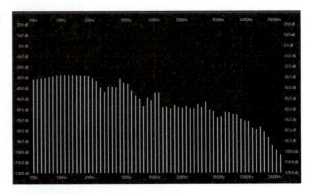

A software spectrum analyzer in action

Often, a special 'reference' microphone is used to obtain a clean signal for the analyzer to read, which helps to avoid the 'coloration' of other types of mics. The reference mic provides a flat signal that doesn't enhance or reduce any frequencies, providing the actual sound of the room or sound that it is reading.

A special reference mic is used with a spectrum analyzer

Dynamic control

As well as EQ we also often need to control signals in other ways to provide a better sound quality and/or give us improved signals to start with. The two main controls are compression and gating (noise gate). Compression enables us to control the levels of signals and if necessary reduce or increase them using the tools it provides while gating enables us to cut out unwanted sounds such as noisy amps or spill from other instruments while still allowing the sounds we want to come through.

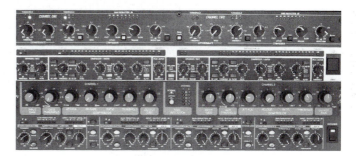

Compressors and gates

Limiting (a type of compression) is also used to 'brick wall' a signal level, preventing it from exceeding a set limit in order to protect equipment or keep the overall level within predefined limits such as a venue's legal sound level. Often, a compressor will include a gate and limiter or they may be combined as part of other processors such as equalizers or management systems.

The compressor

Compressors are often misunderstood. It is easy to use them incorrectly so care should be taken when using a compressor on any signal source. They are often built in pairs and so can be used as a stereo unit (linkable so that the first set of controls operates both channels) or two separate mono units. They are usually used as an insert in the signal chain across one of the mixer's insert points, either on single channels or across buses or main outputs.

A simple example of what we can achieve with compression is if you have a vocalist who sings inconsistently and is quiet in some parts and loud in others then you would have to 'ride' the mixer fader to keep the output sounding good, lowering it on loud parts and bringing it up on quiet parts. If we looked at the singer's performance as a waveform we would see that the loud parts have high peaks or 'transients' and the quiet parts have levels much lower than these transient peaks. If we could hold the peaks down to a fixed level, say reduce them by 3 dB, then we could bring the whole of the level up by that amount (3 dB) making the quiet parts higher in level and reducing (attenuating) the loud parts providing a much more even sound. We have to be a little careful though as we don't want to take away all of the dynamic range or it will sound false and unnatural. Take a look at these before and after compression waveforms to see how it works.

An uncompressed bass guitar wave

The same wave with compression and gain increased. You can see how the peaks are held in check but the lower levels are increased

A compressor will automatically control audio in this way but we have to set the parameters that it will use to work, these are usually as follows:

Threshold Level: This sets the signal level at which the compressor will start to work. If you set it at −20 dB then any signal that exceeds −20 dB will be compressed.

Attack: How quickly the compressor will act after the threshold level has been reached. This allows you to let any short peaks through so that you don't cut off the dynamics of sounds such as the attack of a drum hit. Adding a little time will let them pass through before the compressor kicks in.

Release: This is the amount of time that the compressor stays in operation after the signal has fallen below the threshold level. It enables a smoother and more natural transition than if the compressor just shuts off.

Ratio: This is the ratio of input to output levels that the compressor applies to the signal, so if the ratio is 3:1 then the input signal must rise by 3 dB in order to increase the output level by 1 dB. You may have to concentrate a little here as a lot of minus figures are used for signals. So if the input signal is −7 dB and the threshold level is −10 dB, the input is 3 dB louder than the threshold; got it?

Now if the ratio is 3:1 then the output will be −9 dB as this is 1 dB louder than the threshold level. The input exceeded the threshold by 3 dB but the compressor only allowed the output to increase by 1 dB.

Output Gain: If we compress signals and reduce their dynamic level then we have more 'headroom' so we can increase the overall level of the output using this control.

Compressors may also have a Sidechain or Key Input, which enables them to be triggered by other signals. A common use of this is when a presenter's voice triggers the compressor over background music causing it to dip so that the voice is heard clearly over the reduced level of the music. You may also come across 'frequency conscious' compression where you can set the compressor to act from a certain frequency band. This can be useful for things such as removing 'sibilants' or the 'ss'

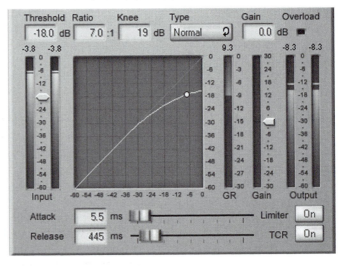

This representation of a compressor shows the input level versus the output level in the form of a graph. The input level is shown horizontally and the output vertically. The white curve is the compression curve that governs the output. The round white marker shows the signal level, which is currently around −8 dB on the input but being compressed to around −18 dB on the output.

and 't' parts of vocals if they are too loud. You would accomplish this by setting the compressor to reduce the level when the offending high frequencies were present.

Some compressors may also have an adjustable 'knee', which is the sharpness of the curve that the compressor uses when it becomes active. A 'hard knee' is a sharp corner whereas a 'soft knee' is a rounded and gradual curve. A hard knee will react faster and apply compression as soon as the threshold is reached but a soft knee will introduce the compression gradually.

There may be other controls but these are the main ones.

Compressor controls including basic gate to eliminate noise

The gate

When backline gear was less than quiet when just running, we used to use 'noise gates' to help cut out the hiss and hum that the backline amplifiers produced. As technology and manufacturing has improved, the noise has become less and so the old noise gates have become simply 'gates'. They can still be used to do the same job but also have many more uses for live sound. The name 'gate' is an apt description as the electronics provide a system that can open and close to allow signals to pass through or not as the case may be. When the gate is open then signals can pass but upon the gate closing they are halted.

They are usually used as an insert in the signal chain across one of the mixer's insert points, generally on single channels although you could place them anywhere they are required.

Gates are often built into compressors and are very useful for eliminating noise that the compressor may increase. If a signal is compressed and then the gain (or level) increased, any noise present will also be increased and can become more noticeable, so 'gating' it out before the compressor helps to eliminate the problem.

A common use for gates is to cut down on unwanted spillage or background noise from other instruments. They can also help to cut out unwanted resonances such as on a drumkit.

Drums are probably the most common use of all for gates and they can be used for controlling or getting creative with sounds. The gated snare sound has become something of a standard for many genres.

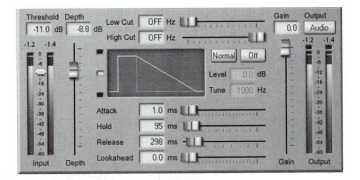

Gate showing envelope in the center box

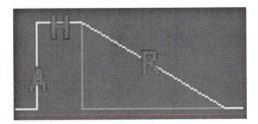

The attack is very short represented by the vertical line A. The hold is then the horizontal line H followed by the slope of the release R (or decay)

Controls may vary but some of the usual ones are:

Threshold Level: This is the level at which the gate opens. If, for instance, it's set at −30 dB any sound above that will open the gate and let signals through.

Depth or Range: This control allows you to set the signal level that is allowed to pass through even when the gate is closed, an alternative to shutting the signal off entirely.

Attack: Controls how quickly the gate opens, too fast a time may cause a 'click'.

Hold: Keeps the gate fully open for a defined time.

Release: The time that the gate will take to close. This allows a smooth transition after the hold time has been reached so that the gate doesn't snap shut audibly.

Output Gain: To adjust the level of the output signal, useful if the processed signal is of a different level to the input signal.

There are usually indicator lights to show when the gate is open or closed, which help in setting up the gate, and there may also be an adjustable knee, the same as the compressor setting.

Some gates also have a sidechain input that allows them to be triggered by another signal (such as a kick drum). This is then used as the source (connected to the sidechain input) to gate the bass, which makes things sound 'tighter' and more together. The bass will be reduced in level after the kick but allowed full power when the kick drum is hit, making them (the kick drum and bass) sound very close in time.

A 'frequency conscious gate' allows you to use built-in 'filters' to define at what frequency the gate becomes active to help reduce false triggering by ambient noise (such as cymbals). A 'key listen' mode can be used to listen to the input and help define the correct frequency.

Two channels of a simple four-channel gate

When setting gates your ears are the best indicator. You will hear if the gate is cutting out any sound that it shouldn't be and you can adjust the controls accordingly. Where you use them is up to you but again don't use them just for the sake of it. If you need to control a sound source then they can be very useful and they also have creative uses (try gating a reverb) but do go easy or you'll end up with a lot of chattering that should sound like a band!

Effects

You probably know a little about the most common effects such as reverb and delay or echo but it's good to know a bit more in depth about them and what they can do.

Effects are often used to add subtle enhancement to sounds, such as adding ambience to 'dry' sounds to make them appear as if they are in a larger space, which is generally more pleasing to the ear. If you're already in a large ambient space then you might wish you could remove some of this but unfortunately we don't yet have the 'ambience vacuum machine'. If I manage to invent it I'll retire to a remote island that I'll purchase with the proceeds and start writing books about lying on beaches.

You can also use effects to add to a sound such as using an echo or delay sound to add repetitions of the original signal in a Rock 'n' Roll style, or modulation effects such as chorus to blend sounds or give them movement. Many effects units contain a variety of effects (sometimes it's possible to use them simultaneously) in a 'multi-effects' unit that can be particularly useful when you aren't able to carry large racks of gear around. They have user-definable presets, so you can save your favorite settings and quickly recall them during the show.

Effects are usually used by inserting them into the mixer's 'Auxiliary Send' system where any channel on the mixer can be 'sent' to the effect. The output of the effect can be brought back to the mixer via a 'Return' input. Often, it will be brought into a channel or pair of channels, allowing the engineer to have the same control as every other channel for the effect, particularly level and EQ. A big plus with this method is that it is quick and easy to pull the faders down on the effect channels when they are not needed, such as between songs when there may be talking, which should be dry.

Each channel will have a send level, so any amount of signal can be sent to the effect from any channel; you could add more effect to background vocals than to lead vocal, for instance. There will be more information on Sends and Returns in Chapter 11, so I'll just get on with the effects themselves.

Effects units often have a wide range of control. Many are digital and use a menu system. They usually have an input level control that should be set when the full complement of signal is going in. If set too high, they may distort audibly, which doesn't sound very pleasant.

Multi-effects units

Delay and echo

An often-used effect is the delay or echo, which basically takes the original sound and repeats it. The classic example is Elvis style 'slapback', which is a repeat after a short delay. It's possible to create a multitude of different effects by adjusting the delay between the repeats, the number of repeats and their level that may be set to diminish in level at a set rate. A most useful control on the

A delay 'Tap' button for setting tempo manually

delay is a 'tap' button that allows you to tap in the rate at which the repeats should occur. This means that you can do it in real time to suit the tempo of the band.

It's best to use delays on single-sound sources although you could use them on groups such as vocals. If you use a very short delay then the repeat can be kept very tight in to the original signal and the effect will be a 'thickening' of the original sound instead of an obvious echo effect.

Reverb

A shortened version of the word 'reverberation', reverb is an ambient effect that consists of many echoes as if the sound has bounced off many surfaces in a real setting or space such as a room or hall. In fact, it's common for reverb units to have settings such as Room, Hall, Cathedral and subsettings of small, medium and large to define these spaces further.

Traditional types of reverb presets are also available, which emulate old style methods of generating reverb such as 'Plate', which was really a metal plate suspended so that it could vibrate when a signal was passed into it. The vibration was then picked up and fed back out, producing an ambient type of sound that is still often favored for vocals. Gated reverb is also an often-found preset that is a favorite for snare drums.

The reverb unit will generally have controls for the length of the decay, room size, input level and a useful predelay setting that allows you to introduce a slight gap between the original sound and the reverb, which helps to make the original sound more intelligible.

Reverbs can be used on most sounds and are good for putting dry sounds into better sounding spaces. If, for instance, you are working in a very acoustically dead space (maybe with lots of drapes and soft furnishings) you can add reverb to the overall mix to make it a little livelier. The reflections give us a sense of distance. Sounds that are some distance away create more reflections, so the more you add the further away the sound will appear to be. If you have a large stage you can use this sense of perception by adding a little more reverb to sounds that are further away and less to those closest, creating a depth to your mix.

If you have the reverb return to your mixer on a channel (or usually two channels for stereo) then you can add EQ and even insert dynamic processing if you like. A bit of high frequency can add a nice sweetness and if you compress the reverb you can get that big '80s' ballad sound, should you want it!

Modulation effects

There are a range of effects that modulate the signal, the most common ones being chorus and flanger. Modulation uses an oscillator to pass the signal content through a waveform (such as a sine or triangle), creating a sense of movement within it. This in simple terms sounds like a swirling that has an adjustable rate (how fast it swirls) and an adjustable depth (how deep the waveform being used is). At low rates and depth the effect can be used to thicken or fatten sounds, giving them a feeling of being bigger than they really are, hence the 'chorus' name. Add some chorus to a few vocals and they'll sound as if there are more singers present.

The flanger effect replicates an old sound that was created when tape was the recording medium and someone decided to hold a finger against the tape reel's flange as it played, causing it to modulate (or flange). Not usually as subtle as chorus but often used on instruments such as acoustic guitars.

A chorus/flanger effect

The phaser is similar but the actual phase of the signal is modulated creating a 'Hendrix' type sound.

There are many other effects available such as pitch shift, tremolo, ring modulation, and so on but these are mostly used as backline or special production effects.

11

Mixing – Consoles and how they work

The Front of House mixer

The mixer is the nerve center of your show and provides all of the main network and controls required to create the 'out-front' sound and often the monitors as well. Every sound that is going to be in the mix will be fed into the mixer, adjusted if required and then sent back out to the PA system amplifiers.

Variously called a console, board or desk, the mixer carries the main level and EQ controls for each channel in a vertical strip and arranged in a 'frame' that usually has many channel strips along with other strips and modules to perform various duties. The most common frame sizes are 16, 24, 32, 40, 48 and 56 although other sizes are available.

Mixers can be analog or digital, with the new generation of digital consoles becoming more popular as time goes by. The new generation of engineers is becoming more accepting of their format and sound than the old analog guys. There are arguments for and against both types. The sound and hands-on capabilities of analog is usually cited as the biggest factor in their favor while digital consoles have the benefit of total recall and being an all-in-one solution. They often contain onboard dynamics and effects processing, obviating the need for numerous outboard racks. Personally I can see the value in both. If I were specifying a mixer for a tour then I could jump either way, with digital looking very attractive. But until a standard for digital consoles is defined (and it may never happen) there's always the possibility of walking up to a digital mixer at a festival and not understanding its system. Not good when you only have minutes to put your show together. At least with analog the controls, in general, will be familiar and easy to get at. This does mean that you may have to learn the nuances of several digital consoles if you don't have the luxury of calling the shots. It's still early days for digital in the live world, so there is plenty of time for change and the larger rental companies usually find their own standard.

Analog mixing console

Digital mixing console

The Monitor mixer

Monitor mixers are not too much different to FOH mixers, so I'm going to cover them both in this chapter simply as mixers. The FOH mixer is often used to provide the monitor mixes as well as the out-front sound on smaller shows where a separate monitor mixer is impractical or too expensive. Monitors often require a large number of separate feeds (mixes), so the monitor mixer generally has a large number of outputs that can provide individual mixes utilizing Aux Send controls from each input channel. This is much the same as an FOH mixer and is achieved in exactly the same way. The dedicated monitor mixer may also have many more output faders to provide fine-level setting for each monitor mix.

Inputs and outputs

The main layout of a mixer consists of mono channels that will accept a single signal per channel. These are usually of the microphone type and consist of XLR sockets on the back of the mixer (although they may also be positioned on top), usually in line with the channel's strip to make identification easy. There may also be ¼" jack type sockets, which can take line signals. In general, the XLR type is most commonly used as they provide a balanced signal that helps to prevent noise. Any unbalanced signals on the stage are converted to balanced by using a DI box. Insert points may be present, which can be on jacks or XLRs. Stereo channels may also be included. These will have a pair of inputs that are controlled by a single-channel strip that controls both of the incoming signals and sends them out to the main mix via a single stereo fader. Sometimes stereo channels may be switched to mono channels.

A multicore snake is used to bring the signals from the stage to the mixer. It may have individual plugs at the mixer end or a large multiway plug that carries all of the connections in a block. At the stage end is usually a box with all of the relevant sockets and connectors required for inputs and outputs. It is basically many sets of wires

Mixer inputs vary

encased in a single outer sheath that replaces numerous microphone cables.

The stage box provides all of the connections required at the stage end of the snake

Digital systems may use a much smaller signal cable that is plugged into converters at each end. The stage end

The 'tails' at the mixer end of the snake

will have Analog to Digital (AD) converters for sending the source signals and also Digital to Analog (DA) converters to receive the signals coming back from the mixer. At the mixer end there will be the opposite, so DA for the incoming sources and AD for the outgoing ones, unless the console is entirely digital in which case there would be no need for conversion from digital at all.

Direct output sockets may be included on some mixers, enabling the signal to be tapped from each channel to feed a recording system or maybe a broadcast system for radio. These outputs may not be affected by the EQ or fader stages of the mixer if they are positioned immediately after the gain stage. This means that any signals taken from this point (post gain) will be affected by any change in the gain control. If they are post-fade then they may be affected by everything including the EQ and fader level. Like many things, this will vary from one manufacturer to another.

Insert points allow you to tap into the signal path in order to process the incoming signal with equipment such as compressors and gates. Better mixers will have a switch so that you can switch the insert on and off and may also have a switch to allow you to use the insert either pre or post EQ. The sockets used are often TRS (Tip, Ring, Sleeve) ¼" jacks that carry both the input and the output. This configuration means that they are unbalanced, so higher-end consoles will have separate sockets for the insert input and output on either balanced jacks or XLRs.

It is also possible to use the 'send' section of an insert to tap into the signal path; as an example you may have a mixer with no direct out sockets, so you could use the insert send to take signals to a multitrack recorder. Depending on the sockets this may mean using a jack plug that is pushed in to the first part of the socket but not fully home. Inserts may be found on individual channels, subgroups or other sections of the mixer depending on its specification.

You should be aware that inserts using TRS sockets are usually wired with the tip being send and the ring being the return. Sometimes this is reversed, which may mean that you have to reverse the connections on your processing gear by swapping the input and output over.

Channel insert sockets may be labeled send/return

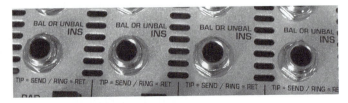

Or INS or insert depending on the mixer

Sockets for other inputs and outputs such as auxiliary sends are usually on ¼" jacks but may be balanced jacks or XLRs on larger consoles.

There will usually be a headphone socket. This may be hidden underneath the armrest and an XLR is often pro-vided for plugging in a 'talkback' microphone so that you can communicate with the stage.

Have a look at Chapter 12 for more information on connectors.

Most of the following information is specifically for analog mixers although the general principles apply to digital mixers as well.

<u>Typical signal flow</u>
<u>through a mixing</u>
<u>console</u>

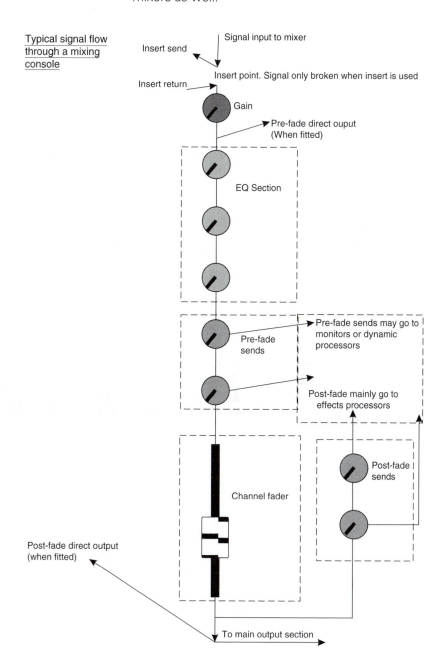

Signal input to mixer

Insert send

Insert point. Signal only broken when insert is used

Insert return

Gain

Pre-fade direct ouput
(When fitted)

EQ Section

Pre-fade sends may go to
monitors or dynamic
processors

Pre-fade
sends

Post-fade mainly go to
effects processors

Channel fader

Post-fade
sends

Post-fade direct output
(when fitted)

To main output section

The channel strip

Although mixers can look extremely complex they consist of a range of modules that are of the same layout and construction. A 32-channel mixer will have 32 channel strips of which most, if not all, will be exactly the same mono strips (the 32 may include some stereo strips that count as 2 channels each). To make life easier we'll look at the main types of strip, or module, separately. Once you understand how the main strips work you will be able to adapt this knowledge for most mixers as they all conform to the same general design.

The mono strip

Mono strips are the ones that you see most of on mixers. They consist of a set of controls arranged in a vertical strip with the signal coming in at the top and exiting at the bottom. There are certain points at which the signal can be split off but for the sake of simplicity we can stick with the top to bottom scenario for now. Mono strips can be used in pairs to control stereo signals as they have Pan controls that send the output to the position you set, anywhere between fully left and fully right. Pan is also used on mono sources to place them in position within the overall mix; traditionally some sounds stay center such as kick drum, bass and lead vocals.

The stereo strip

Similar to the mono strip but this one has two inputs to bring in a stereo pair of signals. These signals can be adjusted simultaneously by its single set of controls. A single fader makes output level adjustments easy and accurate while maintaining the correct balance between both signals. Stereo channels are often used for bringing effects returns back into the mixer as well as sources such as CD players that have a stereo output. The pan control can be used to place the stereo sound or to compensate for a signal imbalance although this is better resolved at source.

Gain and signal controls

The first bunch of controls at the top of the strip handle the incoming signal and prepare it for the journey through the mixer. The gain control sets the level coming in and is sometimes supported by a 'pad' switch that can reduce (attenuate) the signal by a fixed amount, say 20 dB, if it's too hot. It's important that the gain is set correctly to ensure maximum signal with minimum background noise from each source. The gain isn't an alternative volume level, the fader is for that. Gain adjustment is to set the best level coming into the mixer from the source so that subsequent processing is done at high efficiency. If you set the level too low then you may accumulate any noise present in the system with subsequent processing. If too high you run the risk of distortion, so use the meters to get it high on peaks but without going into the red!

There may also be a filter switch that can be used to remove very low frequencies when a source does not need them. These are usually set at frequencies such as 100 Hz or below and are good for reducing microphone stand noise and taking the low-end frequencies out of sources that don't need them, cleaning up their content. High-quality mixers may have adjustable low-cut filters.

A phase switch may also be found here, which will invert the phase of a signal to reduce phasing artifacts when more than one signal is coming from a source. A common use of this is if a snare drum has a mic on top and another on the bottom and they are out of phase (due to the top one picking up the opposite of the sound wave to the bottom one). Hitting the switch on either one will invert the phase (switching it 180°) and, if they are out of phase, will rectify the problem. If in doubt just try it and listen if the sound improves or gets worse.

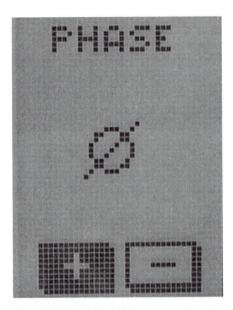

A digital phase switch. Analog mixers may also have a phase switch

The equalization

Every input channel has an equalization section whose complexity and versatility will vary depending on the quality of the console. Budget mixers will have budget EQ with less control and lower quality. Often, a button is provided to switch the EQ in or out, allowing quick A/B comparison with the original sound.

Parametric filters are most often used as they are the most effective for controlling input sources. They consist of controls that allow wide or fine adjustment of frequency ranges allowing you to 'notch' a very narrow troublesome frequency or boost a wider range to improve an instrument's tone. The high and low frequencies may also have a cut or pass filter that will allow frequencies above or below its operating frequency to be cut or boosted by a rotary knob.

In order to understand how these controls work we need to take a look at them in more detail.

EQ sweep controls

Parametric sweep controls may consist of several parts, a frequency control, a cut/boost level control and sometimes a Q (Quality or bandwidth) control. These are rotary knobs that will stop at either end and may have detents at fixed reference points such as the center point of a cut/boost control where no adjustment is made to the level.

Mixer EQ section

The frequency control will cover a range of frequencies in a specified range such as 'Hi-Mid'. You can use it to select the frequency that you want to adjust or sweep it across the range in order to find a frequency for adjustment (more about this later).

The level control can be used to apply changes to the level of the chosen frequency in either direction, cut or boost. It is usually labeled with the amount of gain or attenuation that it can produce such as 16 dB.

The Q or bandwidth control provides a means of defining how wide you would like the EQ filter to be. This means that you could apply a very tight and accurate 'notch' to remove a troublesome frequency without affecting the surrounding frequencies too much.

You could also apply a gentle lift or cut over a wider range perhaps to add more 'bottom end' to an instrument. Some mixers may have a switch to select filter shapes as opposed to a continuously variable rotary control.

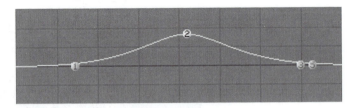

Filter with wide Q of 1.0

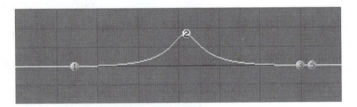

Filter with medium Q of 3.0

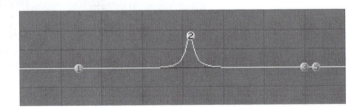

Filter with narrow Q of 12.0

Mixers at the lower end of the price range will often have basic fixed-frequency controls, which don't have sweep facilities but allow simple boost or cut centered at a fixed frequency with a fairly wide bandwidth.

Shelving filters used for the higher and lower frequencies will provide a cut or boost starting around the stated frequency (say 80 Hz). The filter will have a gradual slope defined in decibels (dB) per octave.

Fixed 3-band EQ

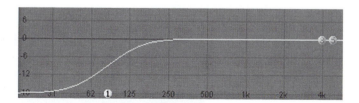

The shape of a low-cut shelving filter

Used in conjunction these controls are very powerful and will cover almost any channel EQ that you require. If you need more it's possible to insert a processor between the source and the mixer (usually at the Insert point) to provide more flexibility.

It's important to understand that making a change in equalization, whether cutting or boosting, can have a marked effect on the gain structure so you need to keep an eye on levels at all times. A reduction in level of a frequency may increase gain if it introduces a phase shift.

Finding a frequency with sweep controls

Sometimes it's difficult to know an exact frequency that you need to control, so the sweep system becomes very useful for finding the right area and then homing in very accurately. Here's how to do it. Select the sweep controls in the correct general area, such as Hi-Mid, and if there's a bandwidth control set it to a fairly wide filter shape.

Now set the frequency control roughly where you think you need to be and increase the level just enough to hear the change. Now sweep the frequency control from left to right until you hear the frequency you are looking for. If you're trying to find out where something is feeding back it will be obvious as the feedback increases and you may need to quickly reduce the level. Once you have found your frequency you can adjust the bandwidth control to a narrower setting, then sweep again (although in a smaller area) to fine-tune around your frequency range. Repeating this will provide a very accurate focus on the exact frequency. If it is a problem frequency then you can now simply rotate the level control into cut to remove its level thus reducing its presence in the channel's output.

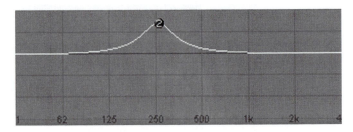

Wide Q boost to find offending frequency

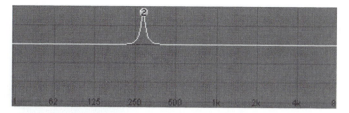

Narrow Q sweep to fine-tune into offending frequency

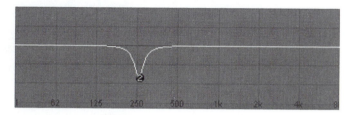

Cutting offending frequency

The sends and returns

Apart from the signals that go out front to the PA speakers it is often desirable to send signals to other places such as monitors or into effects processors. In order to do this we use auxiliary controls (abbreviated to 'aux'), which can split the channel signal and send a copy of it to another part of the mixer. In simple terms it's like having the same signal on another or several channels but they don't usually go through a fader; instead, they are taken through a rotary knob called an 'aux send'.

Mixers may have just one or two aux sends or 32 or more depending on their specification. Sends may be fixed so that they take the signal before the channel fader (pre-fade) or after it (post-fade) or may be switchable. On higher spec mixers they may also be switched pre or post EQ.

Pre-fade sends are usually used where the signal has to remain constant whatever the channel fader is doing, so they're ideal for sending to monitors. If a singer has a pre-fade monitor send then it doesn't matter if his/her level is turned up or down in the PA as he/she will still hear the monitor level originally set up whatever happens.

Channel aux sends

Post-fade sends are good for signals that do need to change with the 'out-front' sound, so they are commonly used for sending to effects units such as reverbs. If the singer's voice is turned up or down then the level of reverb heard from that voice will go up or down with it. If it didn't and the voice was turned down the reverb would sound louder than the voice, which isn't good.

Auxiliary sends will have a master control knob or fader that can be used to set an overall level for the sends, so if your reverb isn't getting enough signal overall you can turn up the aux send master level. An equalization stage may be included for the aux level master controls. The sends will go to outputs on the mixer where they can be taken to wherever they're needed.

Aux send output sockets

Auxiliary returns are the system that can be used to bring signals into the mixer in addition to the channels. They will have input sockets on the mixer and level knobs to control the incoming signals. Switches are usually used to assign them to specific parts of the mixer such as the main outputs or subgroups and they may also have pan controls. They can be used as extra inputs or you could bring your effects back into the mixer through them if you don't have enough channel strips left to use.

The mute and solo system

The channels, subgroups, voltage controlled amplifiers (VCAs) and main outputs will usually have Mute controls so that they can be silenced by simply pressing the relevant mute button. The mute buttons are usually situated close to the fader or knob for the channel or bus that they relate to. Some mixers have a slightly different approach. They may have a button that makes them active (or on), which can be a little confusing if you use different mixers as they may light when switched, making one mixer's lights come on when muted and another's come on when live. Luckily, they're usually green for active and red for muted but can still be confusing. There may be the facility to set up mute groups; these enable the user to select which mutes are applied when a mute group button is pressed. As an example you may have all the vocal channels on a mute group so that you can mute them during instrumental numbers. These systems vary in complexity

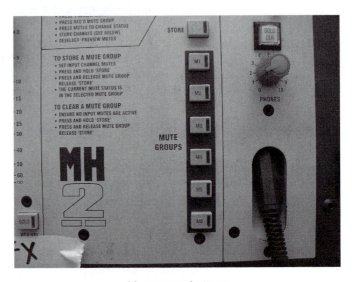

Mute group buttons

and may be simple numbered switches on each channel (numbered the same as the mute group) or may have to be programmed.

Solo buttons are usually situated close to the fader or knob for the channel or bus that they relate to. There may be two modes for soloing: Solo in place and listen modes. These modes are switchable and will make all of the solo buttons become either solo in place or listening mode. Some mixers may have dedicated PFL (pre-fade listen) and AFL (after-fade listen) buttons instead of Solo buttons (see below).

Solo in place will mute all other channels, allowing only the channel with its solo button depressed to be heard. This mode is generally not used too much because if a button is accidentally pressed during a show it can have disastrous results.

Listening modes are called 'PFL' and 'AFL'. PFL is 'pre-fade listen' and allows you to listen to the signal in a channel before it has gone through the fader stage. This is ideal for checking the integrity of sound sources during the show, as it doesn't interrupt the sound reaching the audience or performers. It's also useful for running 'line-checks' when the channels are muted but you need to check if the signals are present before a show. This is often used at large shows and festivals when a full sound-check isn't possible.

AFL is 'after-fade listen' and is used to listen to a signal after it has gone through the full channel or its own part of the system, including its associated fader (or knob). Good for listening to aux sends, especially when used for monitoring, or infill speakers to check if the mix going through the send is correct.

The headphones output on the mixer can be assigned to listen to these PFL and AFL signals, which is the usual method. If you're running monitors from FOH then you can easily listen to any artist's monitor mix by listening to the AFL output from their send. If you want to listen to the drummer's mix and he/she is on aux send 4 then you simply press the solo button located by aux send 4 master control and the mix will appear in your headphones. If you're running monitors then you can do the same thing to put mixes into your listening wedge (Chapter 9).

Solo (PFL) buttons. These are also used as VCA selectors

AFL buttons

Subgroups

As many sound sources come from a group of instruments (such as a drumkit) it's a useful thing to control them as a group when mixing. This makes it easy to turn the whole group up or down without having to adjust every fader in the group. To do this we use Subgroups. The subgroup faders are usually located centrally on the mixer surface. Numbers vary depending on the mixer; 4 or more is usual, with 8 being pretty common on professional mixers. The group faders have Pan controls. If you have a stereo group you can pan a pair of them left and right then any signals coming in will have their pan settings preserved.

Channels are assigned to subgroups using buttons or by programming, so you could pair up subgroups 1 and 2 and then assign all of your drums to them. If you adjust the subgroup it will affect the output from all of the drums. You can still change individual channel settings, so if the hi-hats are too loud you can turn them down (on their channel) without affecting the group. Assigning is something that you should do carefully as it's easy to assign things to the main outputs as well as a subgroup, which may not achieve what you want. You may also have to assign the

Mixer subgroup section

subgroups; so depending on their use they may or may not have to be in the main outputs.

Another use for subgroups is to add processing to a whole group of sources instead of individually such as vocals. This can be particularly useful if you're running monitors from the FOH mixer as any processing inserted into channels will be heard by the performer, but using the subgroup will not affect the monitor sends. I often work with an artist who doesn't like to hear any compression in his monitors but I need to compress his voice for the FOH system, so I put it into a subgroup and compress the group. The processing is added using the subgroup insert sockets. Subgroups can also be used for other things such as sending to different areas of the PA system if they need reinforcing with certain sounds. Often, the center of a large stage may have less vocals, so infill speakers can be placed there and a subgroup could be used to send to the infills via the subgroup direct outputs.

Subgroups can also be used simply to make channels available at a central point. If you have your effects returns at the far end of the mixer then it might be a good idea to route them through a subgroup, which will save you going to them every time a song starts or finishes. You can have everything that you use regularly in one location. There is also a system that simplifies this even more using VCAs.

Voltage controlled amplifiers

VCAs are voltage controlled amplifiers, which may sound like something from an old synthesizer, but they don't (or shouldn't) make any funny noises. Their purpose in life is to control the levels of anything that is assigned to them by using a fader. They don't actually have any sound going through them, so they don't have to be stereo or mono; they simply control the amplifier levels of the sources assigned to them. As an example you could assign all of your drumkit channels to VCA 1. If you move the fader of VCA 1 up or down the level of the whole kit will go up or down with it. This is different from a subgroup as you can't insert any processing across a VCA like you can with a subgroup. A VCA is a single fader without any panning or EQ control, so you can't adjust anything except for level. You can still use subgroups for processing as well as VCAs, so you could group the drums (maybe to compress them) and also control their level using a VCA.

VCAs can be used to control the level of whole chunks of your mix (in fact, all of it if you like) and are very ergonomic on large consoles as they can save you going from one end to the other to adjust things. It's common to put sources such as drums, bass, keyboards, guitars, brass, strings, vocals, effects on to separate VCAs, making for easy mixing. Individual sources such as lead guitar or lead vocal can have their own VCA for fast adjustment.

VCAs can seem a little strange to use but they are very flexible as you can assign any channels to them and preserve the integrity of all the signals. A VCA can control a whole host of channels and their effects, saving you a whole bunch of tweaking when the levels need to change.

You can also assign channels to more than one VCA. As an example you could group things such as drums, guitars and keys on their own VCAs and also have a 'backline' VCA with them all on in case they get too loud.

Mixer VCA section

The matrix

Larger mixers may have a built-in Matrix system to enable flexible routing of signals to various places. The general idea (they do vary) is that sources can be combined and routed to an output. The really useful part of this is that they can be combined with individual level and pan control (similar to aux sends). If you had to feed several areas with sound then a matrix could be used to set up different mixes for different areas to feed infills and balconies for

Mixer matrix section

example. Recording or monitoring mixes could also be set up using a matrix.

Connectivity

The inputs and outputs on a mixer can vary quite a lot and the connection sockets used also vary. This is partly due to costs and partly due to features. More expensive consoles will have better-quality sockets and more of them. At the lower end of the scale, mixers may have fewer XLR mic inputs and a few jack-only input channels with no inserts anywhere. A little higher up in quality and inserts will be of the jack variety using the TRS (Tip, Ring, Sleeve) method where the sleeve of the jack is ground and the tip and ring are used as send and return paths (this varies). At the high end you'll find separate send and return sockets, which will be on balanced jacks or XLRs. Some mixers will also have direct outputs from channels and groups useful for multitrack recording.

If you have to use a different in-house mixer every night when touring then always take plenty of adapters for your FOH rack. Be prepared to repatch your ins and outs to suit various configurations as most venues won't have them.

Processing such as effects will take place using the auxiliary sends going out from the mixer and into the effects unit. This can be returned using channel strips (a panned pair or stereo channel if stereo effects) or auxiliary returns. Channel strips are usually preferable as they provide EQ, can be assigned to subgroups or VCAs and have better control than returns.

Dynamics processing such as compression and gating will use insert sockets either on individual channels or across subgroups.

Main system processing such as house graphic will use the mixer's main output inserts or can be placed in the line between the mixer and the power amps.

Metering

Most mixers have metering of some description whether it's a couple of basic lights to show signal present and signal too high (clipping) or a full set of meters across the mixer in a full 'meterbridge'. Often, there will be a few meters that double up or can be switched. The most common of these is when a PFL or AFL button is pressed a meter will switch to show this level. Clip lights (usually red) show when a signal is approaching distortion and will glow a few decibels before the actual distortion is reached to warn you. In general it's best to get signal levels up around 0 dB when peaking but it's possible to work slightly higher with good-quality mixers. Signal levels can accumulate through the mixer, so a subgroup may have to have its channels adjusted if the group itself is too high signal-wise. Use the meters to get good levels when soundchecking but most of all use your ears. Meters are very useful but you shouldn't mix by what you see, mix by what you hear. Meters will show you if the gain structure is not correct but they won't tell you if the band sound is any good.

Keep an eye on the meters as well as on the stage

Labeling the mixer

Most mixers have 'scribble strips' where they can be labeled with a marker pen but this method is messy at best when it's time to change the labels. Most engineers will use strips of white plastic tape (electrical type) stuck across the scribble strips, which can be easily written on and easily removed. Other strips may be needed next to the aux sends to mark what they are controlling. If you are running monitors from FOH you'll need to label the sends, possibly in more than one place on large mixers. Effects sends should also be labeled in the same way. Any sub-groups or other buses such as matrix sends should also be marked up.

Label your mixer clearly

Abbreviations are required. You could use single letters such as G for guitar or shortened versions such as Guit. Don't forget to label subgroups and VCAs. If you're using in-house effects and processors then it's a good idea to label them as it's easy to forget which gate is for tom 2 when you have a rack of 20.

Snakes and networks

Unless you're working on a very small gig the mixer will have to send and receive its signals over a distance where it is impractical to use separate cable runs. The normal method used is a long run of cable that has many cores running from plugs at one end to a stage box with sockets on the other. These are called 'snakes' or 'multicores'. Larger ones may have big multipin plugs at either end that can be plugged into the mixer quickly and, at the stage

Multipin plugs are often used to connect snakes and cables that carry many signals

end, will have a separate stage box also plugging in to a large connector.

Smaller snakes will have the stage box connected permanently and have 'tails' at the mixer end with the relevant plugs (usually XLRs).

On the stage it's common to extend the stage box using smaller snakes and 'remote' boxes. These will be smaller versions used to carry signals across large stages to prevent the need for running lots of individual cables across the stage. The most common use for them is to connect all of the drum mics into a single box, which is then plugged into the main stage box.

More recent digital systems have been introduced, which consist of converters at each end and very thin data cable that is capable of carrying all of the digital signals with ease. All of the signals are encoded as digital data in the converters and then sent down the cable. At the other end the converters then change the signals back into analog and feed them into the system. The big advantage here is the convenience of the thin cable run. The converters will be expensive initially but the cable is cheap enough to throw away. Due to its low cost it's also easy to use two separate runs so that a backup is always in place should the first cable fail.

12

Connections – Plugs, sockets and cables

The many pieces of equipment used to put a show together almost certainly have to be linked in some way to other components or systems. At present, most of this is done using cables and connectors (some wireless connections are used of course). Luckily, this is one area where there isn't too much of a problem with manufacturers trying to use their own proprietary systems, so something of an industry standard exists.

Globally, mains electricity plugs and sockets may vary but the mains input to the equipment is generally of a similar standard.

Instrument and microphone connections are standardized, as are the connections for mixers, processing gear and speakers.

This chapter is designed to show you the main connector types used and their most popular uses.

Electrical connectors

On large shows the mains supply will often be via a mains distribution system or 'distro', which is a box (or boxes) that has a high current input split into various outlets, usually with safety cutout devices built in. Smaller shows may have a distribution board on the stage or even just standard mains outlets. The country you are in will determine the type of connectors used but the connectors on the equipment end will probably be of a standard type.

A common mains connector for large equipment is the 'Ceeform' connector, which is very hard wearing and locks when connected. It is found in different sizes to suit various loads including 16, 32 and 63 amp.

16 amp Ceeform plug and socket

Most mains input connectors on backline and rack mount gear are of the International Engineering Consortium (IEC) type, which has good connectivity and because of its universal appeal is also easy to replace anywhere. Often called a 'kettle' lead it is also used for many domestic appliances and computers and so it is obtainable from electrical stores that may not have any connection with musical equipment. The plugs are usually molded onto the cable but it is possible to purchase plugs that one can fit by oneself in some countries. As they are cheap it's a good idea to carry at least one spare with you.

IEC plugs

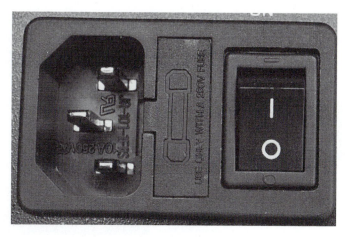

IEC socket

The other common mains connector (although not too common) is the figure eight molded connector. These are generally used on consumer-level products such as CD players and don't have an earth. They are not replaceable plugs.

Special Speakon type connectors are also sometimes used to carry mains but, obviously, care should be taken when using similar types of connectors for signals and electricity.

Older gear may have odd connectors, so it's wise to make sure you can replace them or convert them to the more common IEC type.

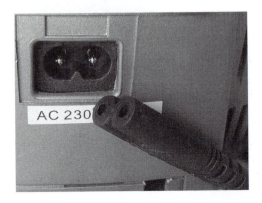

You may find odd power connectors on some equipment

Speaker connectors

Various connectors are used to carry speaker signals and you may well find all of them still in use. Speakers require a fairly hefty cable compared to signal cable, so the connectors tend to get some abuse and as such have to be constructed carefully. If making up your own speaker leads then make sure all connections are good and always use any cord grips to prevent the wires being pulled out.

Jack plugs and sockets are still very common for speakers, particularly smaller, low-powered ones. They are universal and easy to work with provided the cable isn't too thick.

XLRs are also used and are similarly common but may be a little fiddly to work with if cable diameters are large.

Speakon type connectors are currently the best option because they are specifically designed to carry heavy-duty cables and loads. They are easy to fit, requiring no soldering (they have screw terminals), and are readily available in the pro-audio world. Multiple-pin versions can also carry more than one pair of cables allowing a single run of multicored cable to carry signal from several amps to several speakers. The speaker cabinets have input sockets, which then feed the separate signals to output sockets (which use smaller connectors) to carry the separate feeds to each speaker. So you may use a 4-way cable to go to the sub, which then splits the cable internally using one of the pairs to drive its own speakers

Speakon sockets

Speakon plugs

and having a 2-way socket that carries the 2 wires that it hasn't used, i.e. the top cabinet feed. You would then use a 2-way cable to carry the signal from this socket to the top cabinet.

Instrument connectors

The ¼″ (6.3 mm) jack (or phone) plug probably is still the most common connector for instruments. It is available as a straight plug or is right-angled to suit various applications. They usually have to be soldered on but are quite easy to work with and have a strain relief spring or collar as well as crimp-on cable retainers. Mono jacks are used for unbalanced connections.

A variation is the stereo or balanced jack, which has an extra point of contact in the form of a separate ring. Standard jacks have a tip contact and the sleeve but on stereo jacks the sleeve is split into another contact, the ring. Sometimes these jacks are referred to as TRS jacks, Tip, Ring, Sleeve. Stereo jacks may also be used for balanced connections.

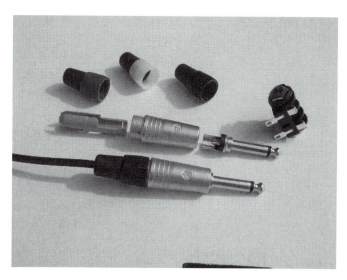

Mono (unbalanced) jack plugs and socket

Stereo (balanced) jack plugs and sockets

The sleeve is usually connected to the cable's screen and the tip to the center core. When a TRS plug is used the tip and ring connections will depend on the application.

Mini jacks are also fairly common, especially on consumer gear such as portable CD and MP3 players. They are very similar to ¼″ jacks but around half the size (1/8″ or 3.5 mm) and are available in straight or right-angled styles.

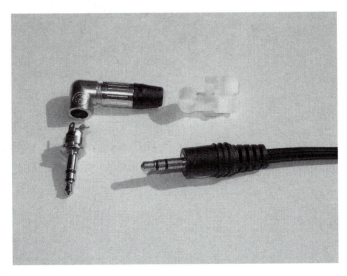

Mini jack plugs

Audio connectors

The XLR or microphone plug is very common in professional rigs as it is the standard connector for microphones, DI boxes and mixers. It uses a 3-pin system (although variations are available) and is very hard wearing. It is mostly seen as a straight plug but right-angled versions are available. XLRs are soldered on and have a good cable retainer and strain relief, usually a rubber boot.

Confusion sometimes arises as to which is male and which is female, as the one with the pins appears to be the receptacle and so could be the female. It is actually the male and the female has sockets that match the pins although it does appear to be inserted into the male shell. The female socket also has a locking mechanism that grips the plug when inserted and is released by simply holding down the protruding tab.

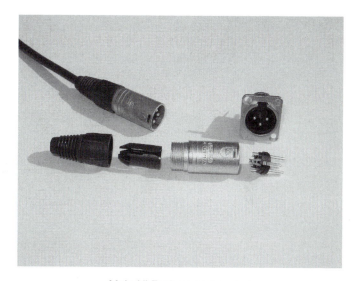

Male XLR plugs and socket

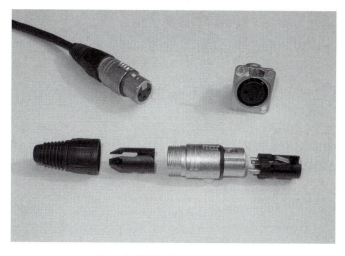

Female XLR plugs and socket

XLRs are very sturdy and the locking mechanism helps to ensure safe connection. They carry balanced signals very well and are the accepted industry standard for connecting balanced inputs and outputs on top-flight equipment.

Jacks are also common audio connectors as are Phono or RCA connectors. Phonos are quite small and often found on consumer equipment such as hi-fi gear. They carry unbalanced signals and can often be found on mixers as tape ins and outs.

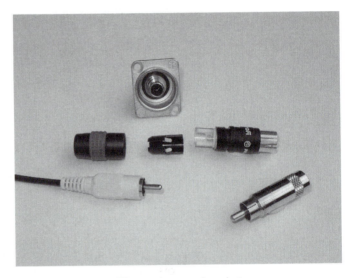

Phono plugs and socket

Musical Instrument Digital Interface (MIDI) connectors

MIDI still mainly uses 5-pin DIN type connectors although only 3 of the pins are actually required. Some systems, such as MIDI guitar pickups, may use connectors with many more pins.

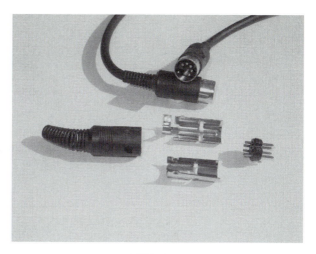

MIDI plugs

Digital connectors

Digital audio technology has used various connection methods with the most popular being ADAT (Alesis Digital Audio Tape) lightpipe, S/PDIF (Sony/Philips Digital Interface) on various connectors, AES/EBU (Audio Engineering Society/European Broadcasting Union) on various connectors and recently Ethernet RJ45 connectors. They are often used in conjunction with AD and DA converters where the analog connectors will plug in.

ADAT was originally the name for the proprietary Alesis Digital Audio Tape recording system. The connections would allow 8 tracks of digital audio to be transferred down a single fiber-optic cable (lightpipe). The recording system has been pretty much overtaken by hard drive and computer recording systems but the ADAT connection format is still popular as it transfers 8 tracks (or channels) at once.

AES/EBU transfers stereo digital audio utilizing a variety of connectors including XLR, RCA, TOSLINK and BNC.

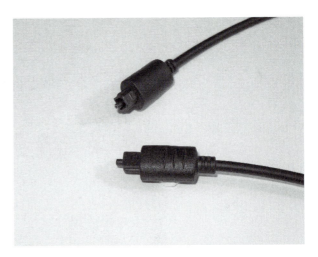

Fiber-optic cable used for ADAT and TOSLINK connections

S/PDIF can transfer stereo digital audio either down a coaxial cable using RCA phono type connectors or down a fiber-optic TOSLINK (Toshiba) cable. These are mainly used in consumer gear.

A fiber-optic/coax (phono) converter for digital signals

Ethernet type digital cables using RJ45 type connectors and CAT5 (Category 5) cable are used to carry many

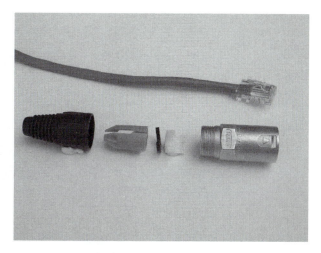

An RJ45 digital connector

channels of digital audio to replace the analog snake cable from FOH to stage.

Cable

Good-quality cable is essential for any audio work in order to retain the integrity of the signals being transferred. Live work also places high physical demands on cables as they are wrapped and unwrapped constantly as well as trodden on and dragged around throughout their life. Failures do happen. It's a good idea to make any broken cables easily identifiable by tying a knot in them or cutting a plug off so that they aren't inadvertently used again until repaired (or replaced).

If possible any tight kinks or bends should be avoided. One of the best methods of wrapping cables up after a show is to coil them by hand, ensuring all twists are taken out, and then wrap a short piece of electrical tape around them to hold the coil together.

Microphone cable is of the two-core screened variety, which carries balanced signals over long distances and rejects outside interference and noise. The two cores carry the signal and the screen shields it from outside interference, this is called 'balanced cable'.

A properly coiled and taped cable

Microphone cable

Instrument cable is usually single-core screened for carrying unbalanced signals and rejecting outside interference and noise. The core carries the signal along with the screen. This is called 'unbalanced cable'.

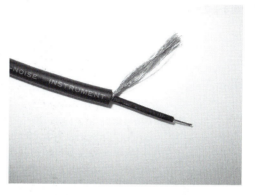

Instrument cable

13

Lighting

Although the main focus of this book covers sound and its associated equipment, the entertainment industry also relies on lighting to make the artists look good, or at least make them visible. Lighting is a specialized subject in itself. There are many fine books specifically about it but as it's such an important part of live events I'm going to cover at least some of the areas that are important.

Theater and concert lighting may differ but the principles used to control the various types of lights are generally the same; in fact, the same equipment is often used with the only variations being the lanterns (or luminaires) themselves.

Basic lighting types

The standard light for rock and pop shows is the Parabolic Aluminized Reflector (PAR) can. They come in various sizes but the largest and standard type used is the PAR 64, or for smaller gigs the PAR 56. PAR 64s use a 1000 W sealed beam lamp while 56s use 300 W versions which are available in various beam widths such as narrow spot, medium flood and wide flood. An alternative to the sealed beam lamp is a system using reflectors and smaller lamps. The reflectors are available in various beam widths and the lamps are usually of lower wattage such as 500 W. This means that more lights can be used if necessary while the overall power consumption is the same. These lamps are much cheaper than the sealed beam types but don't generally last as long and are more fragile.

PAR cans use a frame mounted on the front to carry colored filter gels and may also have 'barn doors' attached to help shape the beam. The focus is fixed by the reflector.

Industry standard PAR 64 cans

As a contrast the much smaller PAR 16 'birdie' lights

Fresnel lights are often used in theaters and have a special lens that produces a wide, soft-edged spread or 'wash' of light. They can be focused by sliding the lamp and reflector assembly in the housing. They may use filter frames and barn doors if required.

Small fresnel type lantern

Floodlights of varying styles are often used, sometimes in large banks (battens) to fill large areas with light.

Floodlighting from behind the band

Spotlights of various types exist and as their name implies they can be focused accurately to light specific areas. The Ellipsoidal Reflector Spotlight (ERS) or 'profile spot' is the most common type and is found in a variety of

Overhead spotlights on trusses

shapes and sizes. As well as extensive focusing the spot may have beam shaping shutters and a place to insert a plate called a 'gobo', which is a stencil used to project a shape or pattern onto the stage.

Follow spots are powerful spotlights that can be focused and moved around to light a performer or various locations on the stage. They are usually controlled by operators but may be controlled remotely using DMX (Digital MultipleX).

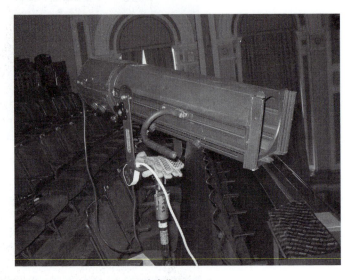

A follow spot

Moving lights, often called intelligent fixtures, can have many and varied characteristics including rotation, tilting, color changing, gobo changing, focusing and strobing. They are popular on large shows and events but not seen as often in smaller shows due to their relatively high cost. They are most often controlled using DMX but may have built-in patterns.

Filters (or gels) are generally thin sheets of special heat-resistant plastic used for changing the color of a light. They are held in a filter frame attached to the front of the light so that its beam shines through them. Other filters are available to perform various duties such as reducing

Moving lights on stage floor

Moving lights behind stage

the output level of a light or making it appear more natural on skin tones.

Filters use a numbering system (e.g. 113 Magenta) but this, like many things, can vary between manufacturers.

It's quite possible to use more than one filter to create a different shade or darken the color used. Some colors may not work well in certain circumstances (such as green on black skin), so be prepared to change.

Filter frames with gels

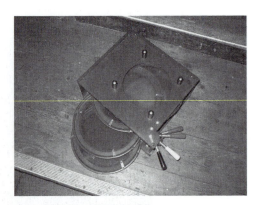

Filters for use with a follow spot

In general, lights are placed above the area to be lit, which is often achieved by hanging them from a truss or bar that is flown above the stage, usually using electrical hoists. Smaller-scale rigs may use stands or tripods that are winched up to position the lighting above the stage. Effect lighting may also include floor lights and lighting that shines up the back of the stage (cyclorama in theaters).

Trusses to carry lighting and also follow spot operators

Lighting fitted above ground level will usually have a safety chain or wire (bond) to prevent it from falling in the event of its bolt fixing failing.

Safety chain on a PAR can

Stage lighting may use various connectors with IEC type and round-pin (15A) connectors being common and often found as the output connector on dimmer packs. Multipin connectors such as the Bulgin plug are also often used, mainly for disco type lights. Multicore mains lighting cables may use heavy duty 'Socapex' style connectors.

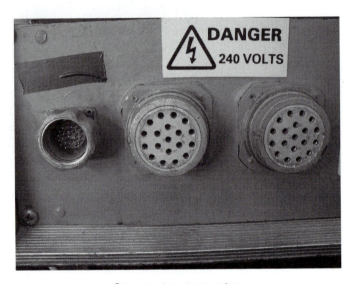

Socapex type connectors

Lighting control

Switching

The most basic form of lighting control is a simple on/off switch. Many small acts can use this method to add lighting at low cost so that they look a little different from the last act that played at the venue. The system could consist of lights just plugged into mains outlets that are turned on and left on during the show or individual switches could be used for each light (or pair of lights). It's possible for artists to control this kind of setup from the stage and there are banks of footswitches available that make it even easier. The drawback with this type of system is its inflexibility (you can only have a light on or off) and the fact that you are switching mains electricity and running mains cables around. Switching mains can cause loud 'bangs' through speakers if the switches aren't correctly suppressed, so it's a good idea to buy properly specified gear. You can also buy switching packs that switch the current when it is at zero voltage (alternating current), thus eliminating the noise. These are usually controlled in the same way as dimmers (using a controller), which we'll look at next.

Dimming

The next logical step after switching a light on and off is to control its level; this is called 'dimming'. Although it is quite possible to use dimmers that use mains power and operate by use of a lever that adjusts the current directly, there is a safer and more commonly used system. The standard across the industry uses a low-powered control system and dimmers that translate the control signals into changes in the mains current.

The controller uses very low power that is safe for the operator but sufficient to carry the signal over long distances. Older systems used around 10 V, which was sent from the dimmer to the controller, then sent back to the dimmer down a multicored signal cable according to the status of the controller. If a fader was pushed up, the signal voltage would reflect this change (while being sent down a specific core) and the dimmer would change the mains current to the lamp concerned, thus making it brighter.

Modern systems use a protocol called 'DMX', short for Digital MultipleX. This system uses digital signals in a similar way but can handle much more information in a smaller cable. The older system had to receive the control voltage from the dimmer and send a return signal down a separate cable core for each channel. If you had 18 channels you'd have to have 19 cores plus common cores (neutrals) and an earth. DMX can operate 512 channels on a single 3-core cable (more of this later).

Dimmers can handle fast flashing as well, so controllers usually incorporate faders and flash buttons for each channel. If you just need to switch lights then a switching pack can be used. This is similar to a dimmer but cannot adjust level, only on or off.

Some dimmers may have a preheat control that allows you to set a level where the lights are gently glowing but not visibly putting out any light. This is useful for keeping the filaments warm to prevent them from blowing due to rapid heating from cold. You may also have the capability to control channels directly from the dimmer or even have a remote control.

Dimmers with 15 amp outlet sockets

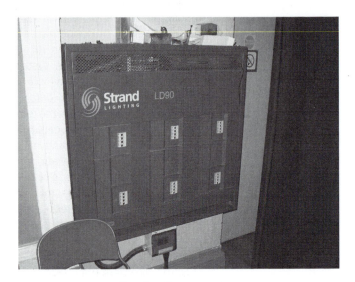

Theater dimmer system

DMX

Digital multiplexing, or DMX for short, is the method most widely used nowadays for controlling lighting of all types. It provides a very flexible system (not unlike MIDI) that can be used to simply switch a light on or to cause a moving light to rotate, change color and strobe at the same time.

Each DMX cable can control up to 512 DMX channels. A console may have several available 512-channel outputs, each of which is called a 'DMX universe'. The data transmitted for each channel can have a value between 0 and 255. This means that a device receiving the data can do different things for each value.

Each light (or device) has an address assigned to it by the user, which can then be used to control it. In its simplest form a light will have a single address, which is usually assigned to the dimmer channel that it is connected to; so the first dimmer pack may have channel numbers one to six, the next dimmer may use numbers seven to twelve and so on. If the operator sends a signal to DMX channel one, then the dimmer will receive it and adjust the light connected to it accordingly. A value of 0 would be off while a value of 255 would be fully on.

Lights and devices that are controlled by DMX are called 'fixtures'. Each fixture has a DMX input and output socket of the XLR type (sometimes five pin). DMX uses cables similar to microphone cables that run from the controlling console and are plugged into the first fixture in the chain. Then the output from that goes to the input of the next fixture along. The final fixture should have a terminator plug fitted or may have a termination facility that can be switched in. It isn't strictly necessary to connect them in the same order as the DMX address order as long as they are all connected. Although the cable is very similar to microphone cable you shouldn't use mic cable as the construction is different. DMX uses twisted pair cable, better at handling any interference that may affect the data than mic cable.

More complex fixtures may have many addresses to operate different attributes they possess such as color changing or rotation.

Addresses are set on the device (fixture) often using increment and decrement controls and a digital display but they may also use DIP switches. Instructions to set base addresses are usually shown on the device or are obtainable in the manual along with a list of available attributes.

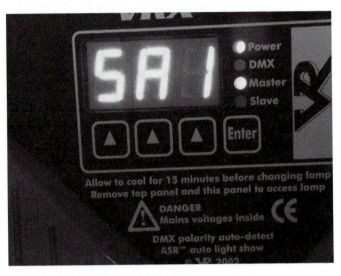

DMX Addresses are set on each fixture

When using DMX it's a good idea to keep a record of all of your fixtures and their attributes close at hand. High-end lighting controllers can be loaded with digital information about fixture attributes to save you a lot of programming. These presets may be already built in or you may have to download them into the console from a disk or computer.

Many of these consoles will also have preset patterns and displays that you can select and modify, again saving hours of programming.

As an example you may have four 'moving head' lights available and you want to do some interesting things with them to complement your other lighting.

Let's say they have five DMX addresses; 1 for on and off, 1 for dimming, 1 for color changing, 1 for pan (left and right rotating) and 1 for tilt (up and down). These addresses aren't fixed, so we need to 'tell' each fixture its starting address. This is the first of the five addresses that it will use, the others will simply be the following consecutive addresses. As an example, if the start address is 27 then the others will be 28, 29, 30 and 31.

Let's suppose we have 48 dimmer channels already running from addresses 1 to 48. We need to use addresses above these numbers and we need 5 'slots' for each fixture. If we start at channel 50 (leaving 49 spare) then we need to use this as the start or base channel for our first fixture (the moving light). We set the address on the light to 50 then the next fixture's start address has to be 5 above it and so will be 55, as the first one uses addresses 50, 51, 52, 53, and 54, that's 5 in total. The second fixture will use addresses 55 to 59, the third will use 60 to 64 inclusive and the fourth will use 65 to 69 inclusive.

Now when we program the console (if it's a programmable one) we must first tell it what fixtures we have, their attributes and their starting addresses. If they have profiles (data about their attributes) available to the console then it's simply a matter of loading in the profile and the console will allocate the rest of the addresses and attributes. If the console doesn't have this facility then you will have to program all of the addresses in and control will depend on the type of console and its features.

The built in 'shape' facilities of some consoles will enable you to select a fixture and choose a shape from

A comprehensive lighting console

Extensive control is available on such consoles

the console's list, such as a circle. The fixture (or fixtures if you've selected more than one) will then perform the shape. You can usually fine-tune the shape by adjusting parameters such as speed and size. This can then be saved as a preset and recalled during the show, often with the facility to modify it in real time. Using these methods you can create many and varied 'looks' that are only dependent on the fixture's attributes and the console's capabilities.

When a console doesn't have such automatic features you need to know the fixture's attributes and the control addresses and values. For example, if color is on the second channel and you want to change it you need to send a signal to that address. The color required will depend on the value sent to that address. If you want green and it's located at values 48 to 63 you need to send a value in this range on the second address of the fixture and the color will change to green. You will probably do this using a fader, so imagine the bottom of the fader as being 0 and the top as 255. This DMX information, or protocol, will be available in the fixture's manual.

A budget lighting console

Special effects

Smoke and pyrotechnics (fireworks) are often used to enhance the look and feel of a show, smoke being particularly easy and effective for making lighting look better. You can purchase a smoke machine quite reasonably and most can be controlled remotely using a special controller or the DMX system. They use special smoke fluid, which is stored in a container onboard, so care should be taken to avoid spillages.

A road-weary smoke machine

There are special systems available for the controlled firing of pyrotechnic devices, which are relatively safe, but as always you must be extremely careful as serious injury can be caused by carelessness or misuse. Cartridges are used for the various effects, which are clean and fairly safe if handled correctly. Firing is via a specially designed detonation control, which usually has a key switch to prevent any unauthorized use. These systems provide all manner of displays from the basic 'flashpot' to large glittering sprays and confetti canons. Specialist companies exist to carry out this sort of work for bigger shows.

Strobe lighting uses high-intensity flashing lights, which can make movement appear to be jerky, like 'stop-frame' movie footage. The only disadvantage with this (and other flashing lights) is that it can cause problems for people who suffer with epilepsy. It is wise (and sometimes the law) to put up signs warning that flashing lights, strobes and pyrotechnics are being used at the entrance to the venue.

Some venues will not allow some effects. It's always a good idea to check before making too many plans or spending money.

Lighting and effects should always complement and add to the show. I have seen some lighting engineers concentrate on a light show and leave the artists in darkness, which isn't quite the idea!

14

Arrival – The venue and setting up

Preparation is everything in many walks of life and live entertainment is no exception. In fact, it depends almost totally on good preparation as there is often little room to maneuver once the show has started, on the part of either the crew or the artist. The preparation starts well before the show when the original idea is hatched and is usually improved as time goes by but each show depends on its own specific setup. The initial arrival at the venue by the crew is when the decisions are made that can make the rest of the setup easier or harder.

Load-in time

The first thing you should do when arriving at a venue is to go inside and introduce yourself to the crew, tell them your name and job and try to remember their names and jobs. You should then ask where all the important things are such as power outlets, equipment, mix and monitor positions, dressing rooms, toilets, showers, and so on. Have a good look around and familiarize yourself with the space and its general feel (including the acoustics) before getting the gear in. You should then have an outline idea about how your show will fit in and work in the venue. Don't rush this. A few minutes spent planning and discussing the options can really make a difference. The local crew will be used to a variety of shows coming in and out, so they should be able to advise you on the best way to work in their venue.

The load-in

Depending on the size and type of venue, the physical loading in of equipment can vary widely. Always try and get some help loading in and use any available equipment such as carts and ramps to make it easier. *Never* struggle with heavy and awkward gear as an injury could put you out of action for a long time.

If you're doing a small bar gig, check everything out first, then decide on the best way to do the load-in. Always make sure someone stays by your vehicle to prevent your gear from being stolen, and work on a relay system for carrying things in. You might just have to carry a few amps, speakers and instruments through the bar, or maybe up some stairs then through the bar. In some cases there may be an elevator to make your load-in easier.

If there are tables and chairs or debris blocking your load-in path, clear them away. Don't risk injury by struggling around obstacles.

If you're working on a large gig and have a truck (or several) then the order of loading is essential and may involve set time-slots for each truck depending on its contents. Lights will usually go in first, then PA, then backline. This means that each team has space to work without encroaching on the other teams too much.

Most of the ergonomics can be worked out before loading the vehicles. Try and pack them so that the gear needed in the venue first is the last to go in the vehicle. As

Unloading is easier with a ramp

an example, if you have a van with backline, lights and PA you should try and put the lights and PA in it first and the backline last. Then when you get to the gig you can take the backline in first, then PA and lights, assuming that you are loading in from the auditorium side of the stage.

Pack your truck sensibly

Another useful but often forgotten tip is to put things where they are going to stay when loading in. I see many bands putting guitars and amps on drum risers during load-in only to have to move them later so that the drums can be set up. Put the amps where they are going to stay and put the drums around the riser leaving it clear for assembly in the center. Guitars can often be put into a dressing room or other safe area until needed. This keeps them out of the way and safe from damage or theft.

Empty flight cases can be a total pain if they're allowed to roll around the stage, so when they're empty, always move them out of everyone's way. You don't have to store them right away but at least don't let them clutter up the stage. The local crew will advise or even move them for you if there's an area set aside for 'empties'. If possible, think ahead about what cases you'll need first and keep them close at hand. Any spares that you might need quickly should also be kept close such as a spare snare drum being kept under the drum riser.

Looking good

Live shows are also a very visual experience for the audience so try and keep everything presentable onstage. If cases have to be used (for instance to lift an amp up) then try and cover them with black cloth. Anything stored onstage should be well hidden and all cables should be tucked away and taped down if running in exposed areas. Have some sympathy for the type of venue that you're in and try and blend in anything possible. If it's a place of worship for instance then there may not be many obvious hiding places for anything, so you may need to use a little ingenuity. Plastic cable ties can be used for holding many things up while gaffer tape is the finest invention known to stage crews and is by far the best thing for securing cables. When you pull the cables back up, always remove the tape, as it's a nightmare of a job trying to remove it if it gets stuck around the cable.

Tape cables down wherever people may walk

If possible, fix cables safely up out of the way

While we're on the subject of gaffer tape it's also useful for a multitude of other things. It comes in several colors, black, silver and white being the most common. The black is great for use where you don't want it to be seen while the white can be used to mark the edges of hazards such as drum risers, which aren't easy to see on a dark stage. Silver is good for general use and also marking spots on stages (such as where a mic stand will be placed). It's usually in rolls that are 2″ (50 mm) wide but 1″ (25 mm) is also sometimes available.

Tape used to mark edges of steps, making them easily visible

Safety is a priority for crew, artists and audience alike and goes hand in hand with tidiness. Proper risk assessment is carried out on larger shows but common sense should be used at all times on any size of show. If you're running a snake out from the stage to the mix position then think about the routing and make it as safe as possible, avoiding any points where it may cause a trip hazard. If cable runs are necessary where people have to walk then use proper cable crossing devices, cover with a piece of carpet or simply tape down.

Onstage cables can usually be run adjacent to amps, monitors and around drums or under risers to keep them out of the way. Place them all as close together as possible. In areas where they have to cross any-one's path you should tape them down and, if pos-sible, use a carpet over them or a proper cable crossing.

A cable crossing

If a crossing point is not easily visible then make it visible using white gaffer tape. You can also put an arrow to show directions using white tape. Try and make them visible to artists but hidden from the audience.

If you have any equipment plugged into mains sockets that are accessible by the general public (or even local crew) then it's a good idea to put a strip of tape across the plug to show that it shouldn't be removed. If necessary a sign can also be placed close by.

When you are all set up and ready to soundcheck, have a quick double-check to make sure that all cases are safely stowed, all cable runs are tidy and secured and any stands (mic, drum, etc.) are properly tightened.

If all is well then you're ready for the soundcheck. Any spare time you have now should probably be spent having some refreshments, as it may be a while before you get another chance.

Safely stow any unused cases or equipment (make sure the fire extinguisher is kept clear!)

Early days

When you first start working on a show or tour it's a good idea to make lots of notes and drawings of how it all goes together. Taking photographs is excellent for keeping a detailed record of things. If you use digital photography you have more storage options. Your photos can be kept on CD, printed out, and kept on your laptop but you could also save them on a web page for easy access from anywhere in the world. Digital photographs can also be easily e-mailed.

If it's OK with the owner of the gear you're using then label everything that you can with electrical and gaffer tape to help remember how it goes together. Drum stands can have their memory locks set or tape stuck on to show their heights, with numbers or letters written on in permanent marker to identify each piece. Leads can be taped together into looms and have each end labeled for quick identification. Small pieces of tape can also be used to write settings on for backline or rack gear and mixers should always have a good labeling system applied.

Photographs of equipment will help when setting it up again

Use tape to mark the position of equipment wherever possible

15

Testing one, two – The soundcheck

When everything is set up and ready to go there should be a soundcheck involving all of the crew and artists that will work on the show. There are occasions when this isn't always necessary, such as running the same show for several nights in the same venue or on fast-changeover festivals. If there is a change in the show, even a small one, another soundcheck should be done.

Soundcheck time

The idea is to get all of the systems involved running correctly and as they will be during the show, or very close to it. An empty venue will sound somewhat different to one with a full audience but we just have to live with that. A priority during soundcheck is to make sure the artists all have a good monitoring set up and are happy with the sound of their monitor mixes. It's much easier to get this right during soundcheck, when any feedback problems onstage can be fixed without the audience hearing them. The out-front sound will also be sorted out and any necessary lighting cues or stage marks can be finalized. If anyone is unhappy with his/her sound then now is the time to address the problem. If you are working on a tour it's a good idea to familiarize yourself with as many aspects of the show as possible just in case you need to help out in other areas. Always pay attention to what's going on during the soundcheck as well as during the show.

The soundcheck may be controlled by the sound engineer who will talk to the artists (often via talkback system or 'comms') or a musical director may run it ensuring

You may want to talk to the artists onstage

things are to his/her liking. However it is done, every artist should be asked if they are happy with their onstage sound before finishing. If you can learn everyone's name then that will get things on a friendly level but if you're unsure then try using the instrument name such as 'is Mr drummer OK with his sound?' or 'how does that sound to Ms lead guitar lady?'

Long-drawn-out soundchecks cause tension and frustration, so always prepare as much as possible before calling artists to the stage. If they drag the soundcheck out then you just have to live with it!

In the time before the artists arrive for soundcheck there are lots of things that can and should be done.

Setting up the Public Address system

Before the soundcheck the PA and monitoring systems will be set up. This usually consists of playing music of high quality through the system and listening for any problems such as too much or too little bass or odd-frequency anomalies. A spectrum analyzer can be used for accurate checking of the room sound but your ears should be the best judge. Using a vocal mic is also a good indicator of the actual sound that will be coming from the speakers, particularly for the monitors, which may have to have EQ applied to reduce frequencies that are prone to feedback. You should make various noises and sounds into the mic to cover a broad range of frequencies. It may sound silly but it works! Sounds such as 'mmm' and 'huh' can be used for low frequencies while 'sss' and 'ch' will provide higher frequencies. The old 'one, two testing' covers a good range especially if the 'one' is said in a low register. You can practice hitting various frequencies using a mic plugged into a spectrum analyzer.

The room itself and the position of speakers will have a marked effect on the overall sound. Experience will show the best way to deal with this but if in doubt try a few things out. If the bass end is boomy then maybe your subs are on the stage causing it to resonate when they would be better on the floor. You may be able to remove harsh upper-mids by repositioning your top cabs. Try it out and see.

You should aim for a good sound before the band arrive. Not easy in this ice rink!

You should use a stereo 31-band graphic equalizer across the PA to correct frequency problems but make sure that the crossover and any other system processing is set correctly first.

Setting up the mixer

Before the band arrives you should allow time to set up the whole PA. I find it's also very useful to get the mixer ready. Use it to set up the PA first with a CD of music that you know well. Try and use music with a similar content to the show and use tracks that contain good frequency content across the range, i.e. bass, middle and treble. Label the mixer and if it isn't set with all controls at their default state, reset it first. *Resetting a mixer back to 'normal' is common courtesy when it isn't your own, so do try and remember to do this after every show.* This gives you a good starting point to work from. Everyone has their own method but here's how I generally set up the mixer before a show:

- Connect all of my processors and effects
- Make sure everything is muted
- Switch on

- Select the subgroups I am going to use and assign channels to subgroups
- Select any VCAs I will use and assign channels to them
- Select phantom power for channels requiring it
- Switch in any low-cut filters required
- If mixer has insert switches, switch on the ones required
- Set up all of the pans for channels and subgroups
- Switch auxes for pre- or post-fade if necessary
- Set auxes roughly to make them easy to spot during soundcheck
- Set basic EQ for channels
- Plug a mic or CD into a channel and check routing assignments work
- Use mic to check if aux sends for monitors are correctly placed onstage
- Use mic to check effects sends and set effects
- Set up any 'extras' such as matrix outputs

Setting up the mixer may take a while

NB: The CD and mic check can help get the system set up but this isn't necessarily written in stone. Always make sure that it sounds good when the artists are on stage regardless of how your CD or mic sounded. The backline may have a marked effect on the overall sound so be prepared to adjust your previous settings.

Facilities vary with different mixers. Sometimes it's necessary to change the connections between my outboard rack to suit the mixer. Inserts can vary in the sockets used (jacks or XLRs) but may also vary when using TRS jacks as some use the Tip as Send and some use the Ring as Send. You may have to check this configuration if things don't work correctly. Flexibility is the key.

If I don't have enough channels to fit my usual setup, decisions have to be made as to what can be left out. It's then time to look at any stereo sources that could be run in mono such as keyboards or drum overheads. After that it can get difficult but some drum mics usually go first, such as tom toms, and overheads will be moved to pick up their sound. Vocals and anything that doesn't have amplification of its own are usually the most important.

If you don't have enough channels then you will have to cut back

Soundchecking the band

There is no fixed way of soundchecking. Often, the order of musicians turning up will dictate the order. A common method is to start with the drums and work from there, taking in the bass and rhythm components, then keyboards, brass, strings, guitars, and so on before finishing with vocals. The mixer, if previously set, will probably be set up in this order, so it is then logical to work across it.

Good signal levels and low noise are prime factors in obtaining a good mix so the input must be at a good level (to help eliminate any noise in the line) and not too high so that it may distort. This is set up using the input gain while the performer plays or sings at show levels. As most artists will increase in level when they are actually performing it's a good idea to allow a little headroom in the gain setting.

Use the PFL button if necessary to check the signal level on the meters and try to hit around 0 dB on peaks. If you go into the red then you will have to back off as you are approaching 'clipping', which will overload the circuitry and sound horrible. Clip lights usually warn you before they reach distortion. It is possible to 'ride' some mixers but I wouldn't advise it, especially on gain stages. Any insert processing such as gates and compressors may be set using the indicator lights at this point.

Once the level is correctly set you can adjust the fader to bring it up into the house. You may want to turn it up quite loud just to see how it sounds at high levels and then adjust the EQ. If at all possible use subtractive EQ and cut unwanted frequencies rather than boosting to correct problem sounds. It is acceptable to boost to improve the sound's tone but always watch your levels, as any EQ change can dramatically change the overall signal level.

As many aux sends are post EQ the EQ should be carried out before adjusting the sends, particularly to monitors as an EQ adjustment could cause feedback in a monitor. You may often tweak EQ during the show but you should try and get pretty close during the soundcheck.

Now adjust the monitor sends while asking any artists receiving the signal if their level is OK and adjust as necessary.

Check the insert processors and how they are affecting the sound in the house.

If possible you can also quickly check your effects levels while doing this. Just use one of your spare hands!

Now move on to the next channel.

Check your effects as you go, if you have time

If you are setting a group of instruments such as drums, have the artist play the whole kit when all his channels are set. I also get them to hit around the toms using single shots so that I can balance pans, gates and levels. Most artists are happy to play a solo.

When all channels are in, you can get the band to play a song. I always like to hear a slow ballad and a rocky number (depending on what they do) and always a song with all of the vocals in (if they have one). The material will decide but you need to hear the quieter and louder songs and also any solo levels.

Vocals are very important and should always take priority over anything else. You won't get many people saying that they couldn't hear the shaker but they will surely tell you if the vocals are getting lost.

If you have subgroups and VCAs then you should be able to set up large groups that can be easily controlled from them, making life easier during the show. You don't have to restrict yourself in the same way as subgroups, so you can group all of the backline onto a VCA if you want to or maybe a brass section and their effects channel.

If there's a problem during soundcheck, never be afraid to stop the band and get it sorted before resuming. It will save time in the long run.

During the soundcheck, at strategic times, ask each artist if everything is OK for him or her, especially monitor-wise. You will get to hear of any onstage problems from them and also show that you are paying attention to their needs.

Try and keep the soundcheck fairly short but always cover everything you need to. If necessary, you can ask one or two artists to hang back after the soundcheck to iron out any specific problems.

Get it right on the soundcheck before the audience turns up

When you're running monitors from FOH as well

If possible, get the monitors 'rung out' using the graphics beforehand to restrict the possibility of feedback, and set rough levels for each performer. I always try and allow a little headroom by winding the monitor well up, then lowering it a little so that when the artist asks for more I can go back up knowing I've got some 'safe' headroom before feedback.

If you have a good technician on the stage then you should be able to work together on this and after a while you'll both be quite fast and efficient. When the band arrives their monitors will be pretty good so they should be happy early on making for a pleasant, and hopefully short, soundcheck.

Tracking down feedback

Keeping an eye on the stage is often the best way of finding out what's feeding back. If the guitarist has just walked up to his mic and started singing then it might well be his mic feeding back, probably in his monitor (especially if he's had his monitor turned up recently). Often, a mic will be moved or may even fall, which can be an obvious problem, but sometimes it can be a reflection occurring from a wall or case. Although microphones are the usual culprits I have been fooled by a guitarist getting too close to his amp and that feeding back (often intentionally), which is then amplified in the PA.

The quickest way to stop feedback is to drop the level. You can then bring it back up and EQ if necessary. It's far better to prevent the problem by careful positioning and thorough soundchecking but things do change. Even temperature and humidity will play a part once the show gets going.

After the soundcheck

When everyone is happy, thank them and tell them you'll see them later. If there are any potential problems deal with them now.

Make sure that everything is muted if possible and if there is an act on before or after you then talk to them about what can be left onstage and what may have to be moved. If you have to move your gear then try and keep it in big pieces so that it's easier to fix back up later. If necessary, mark stage positions with gaffer tape

Any new batteries required should be installed now

If you're the engineer/monitor engineer/backline technician for several acts then you might have to stage-manage
as well if there isn't a stage manager in place. Usually
this means controlling the stage times for the acts and
ensuring they stick to this, as well as to their allotted performance time. You shouldn't really have to do this but
sometimes there may not be anyone around who is experienced enough to handle the job.

16

Good evening – The show

After the soundcheck you'll probably have time to sort out a few things and have some food before the show. Depending on the gig, you might be expected to leave some 'walk-in music' playing. This should just be some fairly quiet music that sets the mood for the show but doesn't spoil it by playing any of the material that will be heard later. Bear in mind that it may run out before the show, so set the player up to repeat or make sure there is enough material time-wise to last.

Make sure your walk-in music doesn't run out

There are often free refreshments available at bigger shows. It's fine to keep stocked up but do be careful to avoid getting into a state where you can't do your job properly. Being drunk is never good for performance from either artists or crew alike.

You should be at your position well before showtime and, if there's an act before you be there before they finish. Check that everything is as you left it or reset it so that it is ready. The stage may have to be cleared and reset for your act, so you need to act swiftly and efficiently. All cables, stands and instruments should be checked to make sure that they are safe and secure. Don't leave anything to chance. If a mic stand looks like it's going to droop down during the show then change it now.

Next you need to do a quick line check, before the show starts

Line checking

The line check is the way to see if everything is still working from the stage down to the mixer (so, in theory, in the PA as well) but without actually feeding it into the PA system.

This is done by playing each instrument or talking into/tapping each mic and checking that the signal is coming into the mixer. You need someone on stage and someone at the mixer to do this. Basically you leave the mute buttons on and check the signal by pressing the PFL button on the channel to be checked. If you like, listen on headphones while someone plays, talks or taps the mic in question. As an example if the lead guitar is on channel 16 put 16 into PFL mode by pressing the PFL button down (it may be labeled 'Solo'), then put on the headphones, making sure they're connected and set to receive the PFL signals. Get the person onstage to tap or scratch the mic on the lead guitar amp as we don't want to actually hear the guitar yet. You should see the relevant mixer meters move and hear the tapping. If not, then there may be a problem in the line.

Use headphones to hear your line-check signals

This should be carried out on all channels where possible. Instruments using DI boxes will have to be played in order for the signal to be registered.

Any problems at this point will usually mean a microphone or cable has failed, so use a process of elimination to find the culprit.

First, swap the mic or DI box. If this doesn't work try changing the lead. If there's an instrument involved it could be that, so try it down another line or make sure it's working in the backline amplifier. I've known bass guitars fail because they have active circuitry and the battery has run down, but if you don't try the bass in another amp then you could easily assume it's the amp that's at fault. Keep calm and work methodically.

Check your connections if a line fails or is noisy

It's better to hold the show up for 10 minutes than to start with a hole in the sound and crew members running around like idiots.

The final countdown

Just before the show begins the instruments should be tuned by the backline technician and the towels and drinks for the band placed in their safe positions. Set lists should be in place for anyone who needs them, and props or costume changes should be ready to go. Radio transmitters should have new batteries and be switched on. Any instruments that the artists will walk on with should be handed to them.

Theaters and some other venues will often make calls to the backstage area announcing how much time is left before the show is due to start. You may get calls such as 10 minutes, 5 minutes and 2 minutes to stage.

A signal should be made from the stage to the FOH position that all is ready onstage and vice versa.

All necessary mutes should now be taken off the mixer. If there's an intro piece (music or sound playback) it should be started along with any smoke and lights required.

The first song or two are the critical times, so full attention should be paid by all crew members. If anything's going to fall over or break it will often be now. The sound will be tweaked to allow for the changes since the soundcheck, including temperature, humidity and artist adrenalin affecting the dynamics.

Be ready for the first number

You should stay focused at all times and watch for any signals from the band, as well as keep an eye on the equipment. It's also good practice to keep eye contact with the other crew members. If you have a communication system keep that in sight in case you are signaled to pick it up (they usually have a flashing light on top).

Anything that presents a danger, such as a drink falling over, should be dealt with immediately. Microphones may fall out of stands and stands may fall over or tilt down if insufficiently tightened. This will affect the sound and performance and should be dealt with right away. If any minor problem is spotted it may be possible to leave it until the intermission rather than run onstage in the middle of the show. Whenever you go onstage to fix a problem there's always the chance that you might bump into or trip over something. Be careful and don't run unless it's really necessary.

The intermission

Many shows will consist of two or more parts (sets or acts) so there will be an interval in which there will probably be a need for some more 'background' music, like the walk-in music before the show. House lights will be put on and the audience may leave the auditorium for a while. This is an opportunity to fix things or sort out any problems encountered in the performance so far. If there is a safety curtain (as in most theaters) then it may have to be lowered, and you might need to ask the lighting engineer to put on a few lights so that you can see on the stage. An intermission will usually last 20 or more minutes, which is ample time to get most things running properly again. All instruments should be tuned and the stage should have a quick inspection to make sure no cables have come unstuck and nothing has fallen over that may present a hazard to the performers. Drinks and towels should be replenished and the stage cleared of personnel before the next set begins.

Some venues will require you to wait for clearance before going onto the stage and will also instruct you when you should leave it.

You might get time to grab some refreshment during the interval

The final curtain

When the show is finished there's a good chance that there may be one or more encores performed, so don't leap onto the stage and start taking things apart until you're sure. If things have been planned properly then everyone will know what the encore numbers will be and will be prepared for them.

This is the time when security has to be pretty good as audience members may get a little excited and want to get onto the stage or backstage. I find that security staff often aren't as observant as they could be. It's always good to be alert and, if necessary, tell security that there is, or may be, a problem developing. I wouldn't recommend getting involved in removing people from the stage but if you have to deal with things yourself don't be aggressive. Just help people off the stage safely and be as friendly as possible. Generally they don't mean any harm and it's just high spirits.

The audience aren't usually supposed to be onstage. These folks were invited

The pirates

A common problem at many live events is people making illegal recordings or filming the show. We all know about 'bootlegs'. They've been around for years but most acts will not want anyone recording or filming their show. You may be required to spot and stop people doing it. Often tickets and posters will state that these things are not allowed but still people will try to do it.

Only authorized filming and recording allowed

It is always preferable to have security staff deal with things, but you may have to deal with things yourself at times. Cameras are fairly easy to spot, especially video cameras with their red recording lights. Their users are often unaware of you if they're peering through the lens. Standing in their field of view and waving at them then making a 'cut' sign (drawing hand across throat) and shaking your head may do the job; they know they shouldn't be doing it anyway. A quiet word that they aren't allowed to film also works well.

Audio recording isn't as easy to spot as microphones can be very small and even in-ear type headphones can have microphones built in.

People recording may well look a little uncomfortable, so if they see you looking at them that may be enough to deter them. Always avoid confrontation if possible. If you do have to talk to anyone make it clear that the venue/band doesn't allow recording/filming and CDs and DVDs are available if they'd like to hear and see more. If merchandise is available at the show you can also point them to the stand.

17

Goodnight – After the show

Once the show has finished, it's time to break everything down, pack it away and load it up. There's usually no time for hanging around, especially on big touring shows, so as soon as possible everyone will get moving on the tear-down.

Clearing the stage

After a hectic show there's a good chance that there may be some debris on the stage. This could easily be a hazard. Make sure that you remove any drinks, spillages, towels, props, and so on first.

Loose instruments such as guitars should be removed quickly as they can easily get damaged if knocked over. In venues where the audience is near the stage it's a good idea to remove microphones and anything close to the front of the stage that may be easily stolen.

All power should be cut from the stage and FOH equipment before it is dismantled.

Packing down the gear

Packing things away can vary depending on the number of people involved. It's a good idea to figure out a logical way if there aren't many of you. Large crews will all have specific jobs, so each will attend to their own duties.

Packing things into cases should always be done tidily and methodically. Often, there's only just enough room in many cases for the contents to fit one way. Cables should be carefully coiled and secured. A common way of securing cable coils is a couple of winds of electrical tape, which can be easily removed on the next gig. Sometimes fabric 'hook and loop' fasteners are used or heavier straps for large cables. Always bear in mind that every case will have to be loaded, transported and then unloaded before it is opened again, so make sure that the contents are safe and secure.

Lighting may have to be last as it could still be quite hot and thus may be fragile and dangerous. If it's flown then it will have to wait until the stage is clear anyway. Instruments and backline could be cased up first, which will leave space for the PA system to be put in. Always try and prepare the gear so that it is lined up in the order that it needs to go out. If your guitar amps can be loaded out first put them on the edge of the stage or in the loading area. Don't just pile things up randomly. Always keep plenty of clearance for people to work and move other things around.

If you have heavy and awkward items then *get someone to help you!* There's nothing macho about being out of work for 6 months because you have a back injury.

Using a rolling riser to move heavy cases

Loading it out

Many hands will make light work, so if you can get some decent help for the load-out then always do. Get your vehicle as close as possible to the loading out door and if necessary use a ramp, carts, dollies, or anything else that will make life easier. Clear the path from the stage to the truck/van so that you can easily take everything out. When loading into a vehicle, bear in mind that anything with wheels can roll around. Tip things over onto their sides when possible to prevent this. If necessary use tie-down straps to secure heavy items.

Make sure there's always someone outside by your vehicles as this is prime time for someone to grab a piece of equipment and disappear into the night. I don't want to sound paranoid about theft or security but it's very easy to forget this aspect. When the star of the show finds out that his/her vintage guitar has been stolen because no one was watching the truck they won't be very pleased. You'll also feel pretty bad about it as you wait for the next plane/train home.

Idiot check

When everything is out and loaded you should do an 'idiot check' all around the venue where there has been any of your gear, including the dressing rooms and FOH position. You do this to see if some idiot has left anything behind or indeed if there are any idiots left behind. If there are two of you around then you should both do an idiot check, just in case. I've known big touring crews to leave things behind and have actually picked up a complete snake for a major touring band that had been left at a large venue by someone not doing a proper check after the show.

If you do lose something then you'll probably be expected to replace it. The lesson can be a tough one to learn.

Merchandising

It's highly likely that there will be some sort of merchandising at the show; you might even be the person selling it. This is another area where security is important as a small box full of CDs is worth quite a lot of money and is very easy for someone to steal. Merchandise can also often be used as rewards or bribes enabling you (with consent of course) to give CDs, T-shirts, and so on to people who are helpful on gigs. An album can smooth the passage over borders, help with that little bit of excess baggage and get your truck loaded quickly, if you're diplomatic about these kind of things.

Merchandise can provide a large amount of income on a tour, so never underestimate its importance. Most crews see it as an annoying part of the job but it can often supplement ticket sales and help keep everyone in a job. Sometimes it can also help to provide a nice little bonus at the end of the tour if everything has gone well, so have some respect for it.

Freebies

Often, there will be spare food and drink just lying around after a gig that was on the band's rider. Usually this is fair game as the local crew will keep it if you don't. I see this as a perk of the job and try to keep a couple of crates and boxes handy for storing it in until I get home. If in doubt, always ask as you might be stealing from tomorrow night's rider!

There may be some spare refreshments after the show

Back on the road

If you are on a bus tour you can now relax and enjoy the ride to the next venue. If you're on a van tour there may be a bit of driving to do. If it's your turn to drive stay alert and be prepared to pull over if you start feeling tired and sleepy. If you're the copilot keep an eye on the driver and maybe chat as you go. Don't take any chances. If the driver starts 'nodding' then tell him/her to stop somewhere safe and take over. If you're both sleepy then stop somewhere and sleep.

Life on the road can be great fun but it isn't worth dying for.

Tour buses have beds!

There and back again – Transport and accommodation

Some people in the industry manage to work locally to their home, particularly if they work at a specific venue. Many others, however, will travel, sometimes globally, working with a touring show or a company that tours with many shows. The method of transportation can vary widely. Here are a few common ways of how gigs and tours can be carried out.

Fly-outs

On a fly-out you will usually send, or take, some gear (but maybe not all of it) to your destination and fly-out on the day before, or even the same day as the gig. Many of these gigs will use rented backline and PA equipment, so often just the essential instruments are taken to keep costs down. This means that you won't have the benefit of gear that you're necessarily familiar with as sometimes it isn't the same as you requested. Even if it is then it won't be set up like your regular rig, so you may have to spend some time preparing it all and getting it to sound right. If you have any data on backup media, this is where it comes into play. Your original notes and photographs made at the start of the tour may also be useful if you have to start from scratch.

You'll usually be chauffeured around on fly-outs unless you are staying in the locality for a while and a vehicle is provided.

Tour bus

The 'classic' tour is the bus and trucks tour where personnel travel on special buses and the gear travels in trucks. A variation of this is when only the backline is taken and can be carried on the bus or in a trailer towed behind it.

The tour bus will be fitted out with beds, lounges, a kitchen, refrigerators, TV, Hi-Fi, video games, surround sound, DVD players, and so on. This all sounds very glamorous but bear in mind that a single toilet is the only form of sanitation and the bus usually drives to the next gig while you're trying to sleep. There isn't much storage space for your clothes and worldly goods and you are living in very close proximity to everyone else on the bus. The beds are often very narrow. There isn't room to sit up in one, so if you're a bit claustrophobic it can feel like you're in a coffin. It's not unusual to find someone has decamped to the lounge when they felt a bit penned in by their bunk.

On the plus side, you don't have to drive and can have a drink after the show. You also wake up at the next venue, if you've slept that is. Bathroom facilities will usually be available at the venue to which you've traveled or arrangements may be made to stop off somewhere suitable for a quick spruce-up. Occasionally, you may have to stop overnight. A location is usually chosen where a 'landline' can be obtained to provide power for all of the bus systems such as refrigeration, heating, lighting and air-conditioning.

It's easy to stay up all night on a tour bus, having a good time and socializing with everyone. It's almost a traveling bar. But don't forget that you will have to unload a truck or two and do a gig when you really want to catch up on the missing sleep the next day. Pace yourself.

Van tour

Van touring can take various forms including:

The band traveling in the van, or following on later
The PA traveling in the van, or a spec being sent out for local supply
Just backline traveling in the van
Crew driving the van
A driver driving the van

And numerous other scenarios besides. Doing a van tour usually means that you'll be staying in hotels. The gear will be readily available when you arrive at the gig as it's in the back of your van. You can often plan the journey to take into account a little sightseeing or to pick up supplies en route. You may be responsible for renting the vehicle, maintaining it and keeping it fuelled up as well as having to drive it. It's always a good idea to have at least two drivers for the van. Someone should *always* stay awake with whoever is driving, day or night. If you start to feel tired and lose focus then pull over and swap drivers right away. Falling asleep at the wheel is easily done and it has killed many people. It just isn't worth the risk. Check the fuel gauge regularly. Take regular breaks on long journeys to refuel the van and the people in it.

You should ensure that the van is suitably fitted out with a substantial divider between yourselves and the gear, enough to stop it sliding forward in the event of a collision. If extra seats are fitted they should be secure and have proper seat belts. These things sound like common sense, and they are, but can be easily overlooked.

Foreign driving

It's possible that you will have to drive in other countries, cross borders and even drive on the other side of the road. If you are going to be doing this then think ahead and check out any local laws, road signage, special equipment that has to be carried, speed limits and any other information you should know before you get there. Motoring organizations and the Internet are good sources of information.

Make sure if you are going abroad that your motoring insurance covers the territories that you are going to. You may need to apply for extra cover for many countries. Breakdown coverage is also something that should be checked when working far away. A minor breakdown could result in a major loss of gigs/money if you don't have decent coverage.

You may need to buy a vignette to drive in some countries

Planning

Just like the show, the journey should be well planned with enough time built in to allow for traffic jams, diversions, road works and other unforeseen problems. It's much better to arrive 2 hours early than 10 minutes late. That 10 minutes will put pressure on everyone for the rest of the day. I've found that a good Global Positioning System (GPS or Satellite Navigation) is absolutely essential and can save tremendous amounts of time. I also carry good old maps. If a problem occurs I dig out the map and use it in conjunction with the GPS to plan a new route. If you do this then you can avoid getting stuck on the same roads as all the other GPS users who will divert and possibly block up the next road along (I've seen it happen).

A GPS system is great for navigation

It may be possible to plan your route stops ahead of time and even check in at the hotel before going to the venue. A passenger should be prepared to go into the venue on arrival and check out the loading-in area, then direct the driver in. If you get lost close to the venue then just stop and ask someone, or call the venue and have them talk you in via the passenger. It isn't always possible to leave the vehicle at the load-in area, so alternative parking may be needed. Just ask someone from the venue staff. They'll probably know just the right place. Bear in mind that you'll want an easy exit for loading out later when you park the vehicle as it's easy to get boxed in by audience cars. Check also if there are any parking restrictions so that you don't end up with a ticket. When parking in a public place you should try and hide anything that identifies the vehicle as belonging to the show (such as that itinerary you left on the dash) and remove any valuables.

Contingency plans

When planning your gigs and tours it's always possible that things might not go exactly as you expect. If you foresee any potential problems, build in some contingency plans. For example, if you have to travel through a town or city where there is a major event taking place (not your gig) then there may be a lot of traffic and general disruption, so a slight detour may be in order.

Equipment-wise there's always the possibility of something breaking down. Spares can save the show, even if they're not up to the usual standard.

Itinerary

When touring you should keep a copy of the itinerary with you at all times. Be careful to keep it safe and away from anyone else's eyes as it wouldn't do for fans to find out where everyone, particularly the artists, is staying. Sometimes the plans will change and you will need to update the itinerary as you go along. Generally, you should check ahead each day to see what's coming up and make allowances for it.

Depending on the territory you're working in, the drives may be quite short, say less than 4 hours. Other times you may be traveling for many hours. You should always prepare for these long runs and make sure that you get plenty of rest and eat and drink properly. It's tiring traveling and you don't want to burn out when you have to concentrate on the show.

Some tours will have random journeys thrown in when a gig has been put in to cover a night off, when a regular gig fell through or there wasn't a gig at all. These are referred to as 'fillers'. The idea is that if there isn't a gig then someone has to pay for everybody to stay out on the road. But if a gig is found then it should be possible to at least pay all of the 'on road' expenses. This is the time when you might end up doing a less than spectacular gig with substandard gear. The whole night you're wondering 'what are we doing here?' You might not be told but you're there in order to keep the tour afloat financially; so, hard as it may be, you'll just have to get on with it.

Hotel rooms and security

When staying in hotels you may be required to give some form of security. This is usually a swipe from a credit card but could even include handing over your passport (not advisable). Any 'extras' you have such as phone calls, pay TV, drinks from the mini-bar, and so on will be deducted from your card. If you decide to throw your TV out of the window or park in the swimming pool then this will probably result in your arrest as well.

One of the most common hotel slipups is forgetting to hand in your key when leaving. Try and get into the habit of asking each other as you get ready to drive off. Many hotels now use the plastic card type keys that aren't as much of a problem if taken. They can also be useful for scraping snow off windshields.

Hotel security can be a worry, so try not to leave anything of value in an unoccupied hotel room. Many will have safe facilities that you can use if you want but they won't usually guarantee total security. If you have small items such as cash then consider hiding it if necessary, under the carpet or on top of a wardrobe maybe.

I always like to leave the room so that it appears as if someone is in it by leaving a light and the TV on. Leave the bathroom light on too and close the door so that it looks like someone just might be in there. Draw the curtains so that the room isn't visible from outside. If your electricity has a switch that is turned on with your key-card then either use another card (the one you forgot to hand in at the last hotel) or fold up a piece of the room service menu and push that in to keep the lights on when you're out.

Never leave any sensitive information anywhere. Always use passwords on computer systems. Be careful not to leave any information on other computers that you use in hotels or Internet cafés.

A handy trick to keep the power on

Communications

Good communications are required at all stages from planning to the actual show, so do try and keep an up-to-date
list of relevant contacts and make sure that your itinerary
has contact details of any personnel that you may need to
speak to. If you're stuck in traffic and know you're going
to be late then a quick call to the venue can make your
arrival easier and shows a good level of professionalism.

Contracts

Although all shows *should* have a contract signed by both
parties beforehand it doesn't always happen, particularly
in small venues. If you are responsible for this side of
business then always make sure that both parties (the
venue/promoter and production company/band) know the
exact details of the contract even if one doesn't appear.
This amounts to a verbal contract and it's difficult for anyone to deny what has been said if it has been clearly
stated on several occasions. I recommend stating your

needs/requirements in a e-mail and printing out that correspondence. It is much better than verbal agreements.

Often you'll find people trying to wiggle out of little things (or even big things) but you must stand firm against them or it will just drag on. Occasionally people deny that they received the tech spec/rider, which they then try and avoid complying with in order to save a bit of cash. A quick check with the records will prove that it has been sent with the contract and as the contract was received, so the rider must have been. A better alternative is to call or e-mail the venue or PA company to make sure they have received it. You can put the rider on your web site and point people to it and/or e-mail it to them. Just try and make sure that it reaches the person who needs to see it. The rider should form part of the contract so that anyone agreeing to the contract also agrees to the rider.

Foreign travel

A nice bonus when working in the entertainment industry is that it can often lead to foreign trips. Some of the countries and places you visit will be great and some not so great but all are an experience that, I think, should be viewed as an opportunity that may never happen again. There are some extra things involved when traveling abroad, which many people will be familiar with. Here is a checklist for foreign travel.

Health

Be sure to take any medication that you need when traveling abroad as it may not be possible to obtain certain items in some countries. Speak to your doctor about any special medication or immunization that may be required for some countries. An Internet search can help to find out any recent problems in particular territories. Basic first aid supplies and medications can provide a good emergency backup and don't take up too much room. Always travel with medications in their original container showing the name, dosage, and the dispensing pharmacist.

Insurance

Travel insurance is essential and not too expensive for a year. In fact, it often costs little more for a year than for a few weeks. Check your policy to make sure it covers the territories that you'll be working in and that the coverage provided is adequate.

You should also have liability insurance to cover you for accidentally causing damage or injury to third parties, particularly the general public. It isn't too costly and some venues require it to a minimum standard or financial level. Insurance brokers will be happy to help.

Vehicles may need special insurance if taken to other territories, so check before leaving. You may also want to insure your own gear. Mention this when you are arranging other insurance as you may be able to get it as part of another policy.

Passports and visas

An obvious but easy to forget item for traveling is your passport. I've seen someone put his passport in the luggage that has gone onto a plane and then, when due to board he didn't have it for identification purposes. Luckily, he had his driver's license and, as this was only an internal flight, it was acceptable. Without his license the plane would have been unloaded while his baggage was checked out. Embarrassing to say the least.

You may also need other documents to travel such as a visa or insurance document. Check well in advance with the embassy of the country you are visiting. Again, the Internet is a great source of information. Try a search such as 'travel requirements for Japan'

A passport and driver's license are always good for identification even when traveling at home.

Currency and cards

Foreign currency may be required if you are staying abroad for some time. It's easy to obtain from travel agents, an exchange bureau or on ferries and planes. Credit and debit cards can be very useful, safe and convenient both for getting cash from machines and for paying directly for things in most countries. You may however have to pay extra charges for foreign transactions.

Language help

Unless you're multilingual chances are you'll come across situations when you can't speak the local language. Learning a few basic phrases will help but often people then assume you're fluent and start talking rapidly in their native tongue! If you're working in the big league you may have a translator on board. If not a phrase book may be your best bet.

Laws and customs

Different cultures are fascinating and often seem strange to us. It's worth bearing in mind that there may be very different customs and laws that we should try and comply

Make sure you obey the law in all countries!

with when working in someone else's country. They will see that you have respect for their way of life and will respect you for it. Foreign laws can be very different from our own, so it's worthwhile behaving wherever you are as the punishment in one country can be much more severe than in another.

Like most of the above, you should check out any relevant information well in advance of your trip.

Wherever you go in the world have fun, work hard and take lots of photos or video so that when you're too old to do it any more you'll have lots of memories to look back on.

Have a good one!

Real-life examples

Gig with no PA system built from existing speakers/amps and digital console with hired-in monitors

When we arrived at the venue there appeared to be just a few speakers and a couple of power amps to drive them, so I thought that I'd better just check that a proper system was coming in. Sure enough, my gut feeling was correct and no system had been booked, the venue thought we would be bringing one. Obviously, someone had screwed up but I didn't have time to get into the politics; I asked to see all of the gear available so that I could assess the situation.

There were a pair of full-range speakers, a pair of small high-range speakers and a pair of stereo amps to drive them; no crossover was in the rack. The venue said that there was a mixer there but no one knew how to use it, so I opened the case to find a digital console of the same type as the one I had in my studio. Obviously, I knew how to use it although I had only used one in the studio but at least it was a start. There was little else, so I had to sit down and work out what to do with almost no budget (they obviously hadn't allowed for hiring in anything). I asked the guy from the venue who was helping me to find some local PA hire companies while I figured out a list. It was a serious hatchet job as I had to halve the input channels to fit the mixer, so the drumkit micing got cut down and stereo keyboards went down to mono. I had to have all of the vocals, so they were my priority.

The mixer had 4 aux sends so I could use these for monitors (although I needed 5 really) but I had no monitors so they were also added to the list, powered being preferable.

I drew up a list of mics, DI's, monitors, snake, stands, cables and EQ's, spoke to the hire company then we hit the road to go and collect the gear.

As time was short I asked the backline tech to set the stage up while I was away so that he could help with the other gear when I returned.

Luckily, the PA hire company were ready when I got there and made sure that all of the gear was prepared in advance. Once back at the venue, the stage was already set up, so I told the backline tech what the plan was and we set about building it up. As I was still one monitor short I used a spare guitar combo amp that I had borrowed from one of the organizers and put an extra mic in the kick drum that was fed into it. This gave me a monitor, which only needed to provide timing to the musician who would otherwise have been without a monitor at all.

I put the graphics inline with the aux outputs and sent their outputs down the snake to the stage where they were plugged into the powered monitors. One of the monitors failed to work, so I had to borrow another small guitar combo to replace it. This was used for the bass player who only needed his vocal in the monitor.

Careful positioning of drum mics and channel use gave me a workable mix and after half an hour soundchecking with the band we had everyone happy onstage and were ready to do a show in plenty of time. I programmed some simple effects into the mixer's onboard system and labeled it so that I could find my way around it quickly.

The only drawback on the show was switching between pages on the mixer, which was only a problem for me and didn't noticeably affect the sound for the band or the audience.

At the end of the show the audience went crazy and so after the encores we packed away and I didn't have to buy a drink for about a week!

All of this took place in Germany and I speak very little German, so I have to thank the people out there for their patience, help and understanding, which enabled me to get the show together.

Rock 'n' Roll festival

This was a festival in a large sports hall involving many bands for over three days so we had some discussion leading up to it while doing other gigs on the tour. There's often time in hotels and at venues to have small meetings, so I had a few meetings with band and crew to make sure that I knew everything that was going on and could pass the information on where necessary. On the day before the gig we had a quick crew meeting and decided that we would go to the venue fairly early in order to assess the show and talk to the on-site crew.

Our showtime was 9 p.m. but we arrived at 2.30 p.m. and the first thing I did was to find the loading area and park in it ready to get our gear in. Next we got our passes from the event manager and I asked where we could store our gear until showtime; an area was available at the side of the stage, so I asked for some crew to help us load in and we put our gear in the designated area. We had to be a little careful to store it safely and also to allow full access as bands were using the area for access to the stage. Next I went and had a chat with the sound engineer, monitor engineer and lighting tech to make sure they had received our specifications and to discuss a few minor changes. The sound engineer was looking after the system for the whole weekend and so had plenty of things set up already and I had arranged with the band to use the supplied drumkit and bass rig, which made life easier for me and the stage crew as they were already set up and wired in. I checked out the dressing rooms for security and after we'd had a good look around we decided to go back to the hotel and have a look around the local town until the evening. I asked security where we could leave our vehicle and parked it before leaving.

After dinner we returned to the venue and checked that everything was OK. The show was now running 15 minutes late, so we had a little extra time to get ready. Our backline tech was new to the business, so I was helping him out setting up the shows and as this was one of his first festivals of this type I wanted to make sure everything went well at the stage end during the changeover. We had to wait until the other band were clear of the stage, so we set up as much of our gear as we could in order that we could just take it on and plug it in. It was set up so that the gear furthest away from us would go on first then we'd work our way back to avoid blocking ourselves in, obvious but easy to forget under pressure. As soon as we got a clear stage we started building and arranging our gear. The band were also on hand to quickly check the positioning of their equipment. Once the stage was set I ran a series of line-checks with the house and monitor engineers to make sure it was all working OK.

Then we were all ready, so I went out to the mix position and had a quick discussion with the house engineer about where all my channels were on the mixer. He'd labeled them up for me to make my life easier and put them in useful groups on VCA faders, enabling easy control of big chunks of sound such as the drumkit.

Pretty soon the announcement was made and the band walked on stage. I had asked the engineer to set up a big reverb for the start of the show where one of the band had to do a big 'voiceover' intro and away we went. I had my set list that I'd asked the band to supply me with and I could then anticipate each song and ask the house engineer to set up anything I needed in advance. It usually takes a song or two to get the mix really tidy in this situation as there are no soundchecks but it soon came together and I even had time to shoot a little video.

After the band finished the stage had to be cleared of our gear right away (no time for backslapping) so I went back to the stage and helped the backline tech remove our equipment which we then broke down, packed away and loaded back into our van. During this load-out I asked for security staff to keep an eye on our vehicle and the gear at the stage end as there were people milling around and it would have been easy for someone to pick up a guitar and take off with it. Half an hour later we were finished and having a well-earned break.

Basic example systems

Small vocal PA

A pair of powered full-range 12″ speakers with horns or tweeters mounted on tripod stands. Small mixer (maybe up to 8 channels).

Small powered monitors could also be used from the aux outputs of the mixer.

Medium band PA

Mixer to suit (maybe 24 channels).
Bass speakers. 15″ or 18″.
Top speakers. 12″ with 1″ or 2″ horns.
Power amps and 2-way crossover.
You could also use bass, mid and top speakers with suitable amps and a 3-way crossover.

Powered mixer PA

Powered mixer or mixer amp.
Pair of full-range speakers.
It may be possible to run separate powered monitors from aux output.

Playback system (for dancers)

Powered full-range speakers would be ideal. Two-way systems with separate powered bass speakers are available for larger shows.

General glossary

A simple overview of terms used in the industry and what they mean. Check out the Theater Glossary for more

AFL: After Fader Listen. Switch that enables the listener to hear a signal in a mixer that is after the fader in its relevant section

Amplifier: Device for making a small signal into a much larger one

Array: A collection or group of speakers arranged together

Attenuate: To reduce the level of a signal

Auxiliary: A system capable of providing separate mixes from the main mix. Consists of 'sends' where a signal is split off from its normal path and 'returns' where a signal can be returned into the mixer. Often shortened to 'aux'

Backstage: Anywhere behind (or to the side of) the stage that is not usually accessible by the general public

Backline: The equipment used by the band or artists usually consisting of amplifiers and instruments such as drums, keyboards, and so on

Balanced: Audio connection that uses two wires for the signal plus an earth shield. Excellent at rejecting noise and retaining signal integrity over a distance

Band pass: A filter that allows a defined range of frequencies to pass through. Frequencies outside the chosen range are attenuated

Bandwidth: The range of frequencies covered by a processing device such as an equalizer

Bar: Metal pole used for hanging lights on. Often a '6-bar' if it has 6 PAR cans on

Bi-amping: Using two amplifiers to drive the low- and high-frequency drivers separately

Bin: A type of speaker enclosure often used for bass speakers

Boost: To increase the signal level. Graphic EQs usually refer to boosting or cutting a signal

Box: Term often used for a speaker cabinet or enclosure as well as its usual meaning

Bridging: A system for running a stereo amplifier in mono and increasing its power output in the process

Bus: A section where a group of signals are brought together such as a subgroup

Cab: Short for cabinet, which is the term for a speaker enclosure

Cam-lok: Mains power connection system commonly used in distribution boxes. Each connection (each phase, neutral and earth) has its own 'twist and lock' connector

Chorus: An effect that modulates the incoming signal often used to thicken vocals so that there appear to be more people singing

Click track: A metronome (usually digitally produced) that is fed to drummers or musicians as a timing guide

Clipping: Signal overload. A mixer often has 'clip' lights to warn of imminent problems

Comb filtering: A modulation type effect. Also a problem that can occur when speakers are not properly aligned, time-wise. The sound from the different speakers will arrive at slightly different times causing the effect. Speaker position and delay times can be used to overcome the problem

Combo: Combination speaker cabinet with amplifier built in, e.g. a guitar combo

Compressor: Processor for controlling the dynamic range of a signal. Simply put, it can restrict high peaks (transients), which enables the overall level to be controlled easily. Generally used to keep loud signals in check and increase the level of quiet ones to make them more audible. Often includes a noise gate as any noise present may also be increased

Crossover: Device used for splitting frequencies into ranges such as low, mid and high. The ranges are then fed to drivers that are capable of handling them efficiently. This also serves to protect them by avoiding damaging frequencies entering drivers that can't handle them

Delay: Device used to reproduce the incoming signal at a specified time after the original has passed through it. Usually used as an echo type effect. Also a processor that is used to time-delay the main PA sound to speakers that are placed further back from the main PA so that they sound 'in-time' with it

DI box: Direct Injection (or Input) box that allows unbalanced signals, usually with jack plugs, to send signal down a balanced (mic) cable. Available as Active or Passive versions; Active types require a battery or can use 'Phantom Power' sent from a mixer (+48 volts). They often have an earth-lift switch to eliminate noise caused by ground loops and may also have a 'pad' or two to cut down the level of strong input signals (such as a synth). Often used for bass guitar, keyboards and many acoustic instruments with built-in pickups. The player plugs the instrument's jack lead into the DI box and then connects another jack lead from the box to their amp (if used). Acoustic instruments will often use the monitoring system to hear their instrument if they don't use an amp

Downstage: The front of the stage closest to the audience

Driver: General term for a speaker or horn

Drumfill: Monitor speakers for the drummer. Could be a single wedge or anything up to a pair of stacks. Often includes a sub so that the drummer can 'feel' the kick drum

Earth-lift: Usually found on DI boxes but also on other gear; it's used to switch the earth (or ground) out of the signal path to remove noise caused by ground-loops

Empties: Empty flight cases that need to be stored during the show

Feedback: Noise caused when a signal is in a loop that is continually amplified. A microphone picking up the sound from a speaker which is then amplified again, fed back into the speaker and picked up by the mic again ad infinitum is a good example

Filler: A gig that's put into a tour to cover costs. May not be very good but it will help to keep everyone on the road. Could also be a song added to a set or album to pad it out time-wise

Filter: A device used for selectively allowing or restricting the passage of audio such as frequency content for equalization

Flight case: Heavy-duty box or container that protects equipment. Usually with metal corners and edgings. May have castors

Flown: To fly something is to hang it from the ceiling or other structure. Speakers may be flown above the audience for better coverage

FOH: Front of House. Usually means where the mixer is situated in the main auditorium but can also refer to anywhere in this area or out to the foyer, box office, and so on

Foldback: The monitoring system usually consisting of speakers, amps and control that point back into the stage so that the performers can hear themselves and anything else that they need to

Gaffer tape: Multipurpose hardwearing tape mainly used for fixing cables down. Usually 2″ (50 mm) wide

Gain: The amplification of a signal. Also a control on a mixing console for preamplifying a signal before it reaches the processing inside the mixer (EQ, etc.)

Gate: See Noise gate

Graphic equalizer: Dynamic processor that enables the user to boost or cut selected frequencies in a controlled manner. Often referred to simply as a 'graphic'

Ground: The connection to earth (literally) as a safety device. Commonly referred to as 'earth'. It's best to use a common grounding point for all audio gear as several points can introduce noise in the form of 'ground-hum'

Ground-lift: A switch for cutting (or moving) the grounding point on a piece of equipment to eliminate ground loops and thus hum. You should never remove the earth from a mains power cable or connector

Head: Amplifier that is placed on top of a speaker cabinet such as a guitar stack comprising a head and two cabs

Horn: Special type of speaker component generally used for higher frequencies. Often used as a term for the complete horn assembly but is specifically the 'flare' section, which is generally plastic or aluminum. Can also refer to the shape of a cabinet

Infill: Extra speakers added to PA system to provide sound for areas that aren't covered well by the main system such as close to the stage at the center or balconies

IEC: Very common mains power connector (International Engineering Consortium)

Impedance: Resistance of a device measured in ohms

Insert: A socket on a mixing console (may be on other pieces of equipment too) where it is possible to break the signal path to send it through a processor such as a compressor. Higher-end mixers will have separate send and return sockets across the insert point

Isolation transformer: A transformer that passes signal without a physical connection. Great for eliminating ground noise

Jack: A connecting plug, most commonly refers to the ¼″ (6.3 mm) jack plug

Kilohertz: 1000 Hz (cycles per second). Abbreviated to kHz

Line-array: Specially designed speaker system that is physically small but has wide dispersion. May be stacked or flown

Limiter: Processor that restricts the signal level usually to protect equipment further down the line

Matrix: A system found on some mixers that allows signals to be grouped and sent to dedicated outputs. May be used for a variety of purposes such as recording or for infill speakers

Meat Rack: A frame, usually on wheels, for easy transportation of lighting already fixed to bars (or trusses).

Mic or mics: Abbreviation for microphones

MIDI: Musical Instrument Digital Interface. System that uses digital messages to communicate between various pieces of equipment. Most commonly for keyboards but can be used for many devices

Mixer: Variously called a mixing board, desk or console, it is the heart of a live PA system where all of the signals are brought together, corrected, enhanced and processed as necessary before being added together in what should be a pleasing way to the listener. Facilities can extend beyond this for creating feeds to other areas and recording machines, and so on. They are also used to provide monitor mixes for artists while smaller versions can be used to submix any type of sound such as a bank of keyboards

Monitors: Speakers used for direct listening on stage by artists. They may also be speakers used to listen from a remote location such as an outside truck

Monitor Engineer: Mixes the sound that the artists hear onstage, usually from a position at the side of the stage. May control many different mixes (could be one for each artist or group of artists). Does not respond well to people shouting!

Noise: Unwanted signal produced by equipment or ambient sound

Noise gate: Processor for controlling the level of unwanted noise in a signal path. Often just called a 'gate'

Overhead mic: A microphone usually on a stand that points down toward the instrument, often a drumkit, but could be strings or other ambient instruments. Often used in stereo pairs

PA: Public Address system. The big speakers used to make the act heard in the venue. The term also covers the associated amplifiers, mixer and peripherals

PFL: Pre Fader Listen. Switch that enables the listener to hear a signal in a mixer that is before the fader in its relevant section (usually the input channel strip)

Piezo: A type of tweeter that uses a piezo crystal to produce its sound

Pre-amp: An amplifier used to increase signal to a useable level such as inside a mixer

Raked Stage: A stage that slopes toward the audience in order to help visibility from the auditorium. Can be slightly awkward for some musicians such as drummers; sometimes risers are available, which sit level on the stage

Riser: Platform used to raise performers and their gear above stage level. Often used for drums, percussion and keyboards. May have locking wheels for quick movement as in a rolling riser

Roadie: Member of the crew. Used as a general term for crew but specifically refers to general crew who don't have a specific job

Send: Part of a mixer where signal can be split and sent to a separate output while still remaining complete in its normal path

Set List: A list of songs or pieces that are to be performed by the artists. Should be updated if changes are made. Useful for writing cues on

Sidefill: Monitor speakers placed at the sides of the stage, usually only on bigger stages such as festivals and large shows

Snake: Multicore cable that runs from FOH to stage and carries all necessary signals to and from the mixer to the stage. Usually terminates in a stage box that has sockets for microphones and DI boxes to plug into along with main signal returns from mixer such as PA feeds and monitor sends if necessary

Sound Engineer: The person who mixes the band's sound. May also control the monitors from the FOH mix position

Soundcheck: The process carried out before the show when the venue is still closed. Enables the sound and lighting to be finalized with the artists and their monitor mixes to be set

Spectrum analyzer: A system that reads the sound frequencies in a room via a reference microphone and shows a real-time visual representation of the frequencies and their levels. Can be used to help set up a PA system

Stage Left: Your left if standing on the stage looking out into the auditorium, the audience's right

Stage Right: Your right if standing on the stage looking out into the auditorium, the audience's left

Stage Relay: Often used in theatres consisting of microphones over the stage that can be heard in the backstage and dressing room areas. Has been the downfall of crew members who can be heard commenting about acts when working on the stage

Sub: Bass speaker (subwoofer) or bass frequencies usually below 150 Hz

Talkback: System for talking to the stage. May be a network in itself or simply a microphone plugged into the mixer and sent to the monitors

Technician: Skilled crew member usually specializing in one area, e.g. backline technician or system technician. Usually abbreviated to 'tech'

Throne: Usually the drummers seat as in 'drum throne' or could mean the toilet as in 'he's on the throne'

Top: High speaker or high frequencies, treble. 'Taking a bit of top off' would mean reducing the high frequencies (treble)

Transducer: Any device that receives a signal and changes it into a different type of signal. A microphone receives sound and changes it into electrical signal while a speaker does the opposite. They are both transducers

Traps: Slang term for drum hardware (stands, etc.). These are kept in a 'traps case'

Trim: Another name for the Gain control on a mixer

TRS: Tip, Ring, Sleeve. ¼" jack plug used for stereo or balanced connections

Truss: Metalwork, usually tubular, for holding lighting and sometimes speakers above stages. Can also be used vertically at the sides and may be a very large construction

Tweeter: Very small driver used to reproduce very high frequencies

Unbalanced: Audio connection that uses one wire for the signal plus an earth shield. Used for most instruments and short cable runs. Does not reject noise as well as balanced connections

Upstage: The rear of the stage, farthest from the audience. Also a term used when an artist goes one better than another artist as in 'he upstaged her'

Wedge: A wedge-shaped monitor speaker that points toward the musician so that he can hear himself and any other instruments required. The wedge shape offers a good angle to project the sound from and also a low-profile onstage.

Wireless: Such as a 'wireless mic' that operates by transmitting radio frequencies to a receiver. Wireless monitors (in-ears) work the opposite way as they are the receiver. The transmitter is connected to the monitor mixer and sends the signals via radio frequencies.

Woofer: General term for a speaker, as opposed to a horn or tweeter

XLR: A plug or socket connector usually used on microphones and professional equipment. Usually 3 pin but other variations are available for other uses

Theater glossary

Because theater terms are specific to their own world a separate glossary is required. Some of the words do cross over with other types of shows but some are specific and have different meanings in theater.

Advance bar: Lighting bar hung from the auditorium ceiling as opposed to over the stage to provide lighting from the front

Ambient light: General indirect light from reflections or non-specific light sources

Anchor: Fixing to stage floor, e.g. scenery

Animation disc: Rotated in front of a light to create moving effects such as clouds. May be called an 'animated gobo'

Apron: Part of the stage projecting into the main auditorium in front of the curtain

Areas: Places that are defined lighting-wise. As an example there may be an upstage left area where certain actions take place

Back light: Lighting from behind scenery or actors for dramatic effect such as silhouette

Backdrop: Usually a cloth-based hanging at the rear of the stage. May be called a 'black' or may be painted as scenery

Backing: Scenery used behind an opening in the main scenery such as a window or a door

Balcony rail: Lighting bar fixed to the balcony

Bar: Tube, usually suspended but may be stage mounted vertically, for holding lighting

Barn door: Fitting for front of lights to help shape the beam. Usually has two or four metal plates (doors) that can be pivoted

Barrel: A bar suspended on lines to hang scenery on, so that it may be lifted in and out

Batten: Length of wood or metal tube used in scenery construction

Blackout: Turning off of all lighting to darkness. There is often a button or switch on lighting controllers to perform this in one hit

Blacks: Black cloth used as backdrop and to drape where necessary on stage, often to hide unsightly items such as cases

Bleachers: Moveable seating that can be stowed away when not required. Often electrically powered

Book flat: A pair of flats hinged like a book so that they can stand up

Border: Masking above the stage to hide lighting and so on

Box set: A series of flats usually used to make a 'room' scene

Brace weight: Weight used to hold scenery

Brail line: A line used to hold a suspended item such as scenery in a set position

Bridge: Structure over stage to allow access to lighting

Canvas: Cloth often used to cover flats

Call: Directions to performers of time to start of show, such as 15-minute and 5-minute call

Catwalk: Structure over stage to allow access to both sides without going across the stage

Center line: The line running down the center of the stage. Sometimes but not always marked

Center stage: The middle of the stage

Chase: Light sequence moving from one light to another usually programmed on lighting console

Cleat: Fixing around which lines can be wound and tied off

Color call: List of colored gels required for a show

Color changer: Remotely controlled device mounted in front of a light that can be used to change its color

Color correction filter: A filter or gel used to change the color temperature of a light so as to change its attributes such as providing a warm or cold look

Color wash: A widely dispersed colored lighting look

Counterweights: Weights used to balance scenery or objects being flown on a pulley system. Makes lifting and lowering much easier

Crossover: Backstage passage allowing crossing from one side of the stage to the other without crossing the stage. Also part of speaker electronics

Crossfade: The fading up of lighting while fading others down. Some controllers can perform 'dipless' cross-fades which help avoid uneven lighting or 'dips'

Cue: Signal for an action to take place. May be for performers or crew. Cues may be given verbally via intercom systems, by cue lights or by a written cue sheet

Curtain: Stage drapes and also a theater term as in 'one minute to curtain up' meaning the curtain will open in one minute thus signaling the start of the show

Curtain call: When the actors return to the stage to take a bow and receive applause

Curtain line: The line across the front of the stage where the curtain rests when closed. This should be kept clear especially if there is a heavy steel safety curtain (or iron) which may be dropped in

Cyclorama: The rear wall of the stage, usually curved. Called cyc (sike) for short

Dark: When the theater is closed it is referred to as dark

Dead: The final level of a suspended object (such as scenery) that may be raised or lowered into position. Dead hung is a fixed position of an object (such as lighting), which will not be moved

Diffusion filter: A filter used in front of lighting to diffuse the light. It can provide a softer more even light. A popular type is the 'frost' filter

Dimmer: Electronic device for controlling the level of attached lighting. Usually has several (or many) channels and is operated remotely with a lighting console

Dips: Recessed boxes in the stage containing sockets. May be for electricity or signals

Dock: The scene dock is where scenery is stored next to the stage. The loading dock is where items are loaded in and out, usually close to the stage

Downstage: Area of the stage closest to the audience

Drapes: Curtains and fabric hangings/coverings used onstage

Drencher: A fire safety sprinkler system

Dry ice: Frozen carbon dioxide (CO_2). Used to create low-lying fog by immersing in hot water. Requires very careful handling

Dutchman: Cloth used to cover gaps in scenery

ERS: Ellipsoidal Reflector Spotlight. Hard-edged light

Fade: Dimming of the lights such as a 'fade to black'

Fill: Light used to cover dark or shady area

Fire curtain: The safety curtain

First electric: First lighting bar behind the proscenium arch

Flat: Scenery made from cloth fixed to a wooden frame and painted

Flipper: Hinged piece of scenery

Flown: Object hung or suspended such as speakers or scenery

Fly: To lift up or suspend objects such as scenery

Fly gallery: Area where control of flying gear is carried out. Usually at the side of, and above, the stage

Fly loft: The space above the stage where scenery is hoisted up out of the way when not being used. Sometimes called the 'flies'

Flying iron: Metal device used to fly equipment

Flyman: The person employed to fly things

Focus: Aiming and setting of lighting positions. The lighting engineer will focus all lights before a show

Follow spot: A (usually large) spotlight operated by hand. The operator 'follows' the artist around the stage with it

Footlights: Lighting along the front of the stage pointing upstage

Forestage: Area in front of the stage curtain

French brace: Hinged bracket fixed to flats (or scenery) to hold it up

Fresnel: Soft-edged light with adjustable beam width. Adjustment is achieved by sliding the lamp forwards or backwards within the light

Front cloth: Scenery placed close to the front of the stage, usually to hide scene changes behind

Front of house: In theater, this generally means anything in the audience areas including front of house staff such as ushers

Fuzz light: Revolving or flashing 'emergency'-type light

Gate: Abbreviation for a noise gate. Opening in a spotlight to allow insertion of gobos or other fittings

Gauze: Mesh fabric screen for special stage effects. Can be used with actors behind and specially lit for silhouette or shadow looks

Gel: Clear plastic sheet placed in front of lighting to change its color

Ghost light: Light left on in theater when it is empty

Gobo: A template like a stencil, usually metal, placed inside a light to project a shape or pattern (such as a star or clouds)

Grand Master: Main fader controlling lighting console's overall output

Green room: Backstage room used as a waiting/relaxation area by artists

Grid: Framework over the stage that supports flying gear, pulleys, and so on

Grommets: Metal rings used to hold lines

Groundrow: Low-lying scenery (such as landscaping). Also the name sometimes used for a row of floor-level lighting

Hanging: Fixing a light into position

Hanging iron: Flat metal hook used to hang scenery

Hemp: A natural rope used for hoisting scenery

House: Main part of the auditorium where audience sit

House lights: The main auditorium lighting

House tabs: Main stage curtains (house curtain)

Hook up: Lighting plan including connections

Hook clamp: Standard clamp for hanging lights

Intermission: The interval between acts

Iris: Device inside a light used to vary its aperture diameter and thus the size of the circle of light it projects

Lamp: Bulb inside a light

Lantern: Commonly used term for a stage light

Leg: Narrow curtains at sides of stage. The same as tormentors

Light curtain: Row of lights fixed together

Lighting board: The lighting control console

Limes: Old term for lime lights. Sometimes refers to placement of follow spots

Lines: Ropes or wires used for suspending and hoisting scenery

Load-in: The loading in of equipment for the show

Load-out: The loading out of equipment after the show has finished

Loading dock: Area where load-in and load-out takes place

Loading gallery: Area above the fly gallery for fitting scenery counterbalance weights

Louvers: Ceiling with openings which allow lights to be fitted and shone through

Masking: Scenery used to hide areas from the audience

Muslin: Fabric used in construction of scenery and flats

Offstage: Off the main performance area

Onstage: On the main performance area

OP: Opposite Prompt side of the stage depending on where the prompt is

Packing rail: A tubular rail where flats are stored

Paper tech: Discussion of show technical requirements and their cues, and so on

Pass door: Door from front of house to backstage

Patch: Connection. Electrical or signal. May be made at a patchbay or panel

Pebble convex: Lens type with defined edge

Pin rail: Rail used to which lines are attached. May be in the fly loft or at the side of stage

Pit: Area in front of the stage where the orchestra reside

Plaster line: Imaginary line on upstage side of the proscenium

Profile spot: Spotlight with beam-shaping capability

Prompt: Person who follows script and reminds actors if they forget lines

Prompt script: Book with all documents pertaining to the running of the show such as script and cues

Props: Objects used in a show. Set props include furniture, vehicles, and so on. Hand props may be weapons, newspaper, and so on

Proscenium arch: The archway (or position of) at the front of the stage. It frames the stage and performance and is often ornate

Purchase line: Line used by the operator to move scenery

Rake: Angle of the stage. A raked stage slopes toward the audience

Raking piece: A wedge to level an object placed on a raked stage

Read through: First run through the script

Return: Small flat fixed to a larger flat at an angle

Revolve: A circular revolving stage

Rig: To set up the scenery, lights, and so on

Rope lock: Device for locking off a rope such as a counterweight rope

Rostrum: A platform

Run through: A complete rehearsal

Runner: Length of stage that can be removed to raise scenery

Safety chain: Chain (or wire) to prevent a light or other object falling if its normal fixing fails

Safety curtain: The fireproof curtain at the front of the stage to prevent the spread of fire. Usually steel and with a device to allow it to drop in quickly. Often called an 'iron'

Sandbag: Fabric bag filled with sand used as a weight or counterbalance

Scene pack: Scenery for a particular set

Scrim: Cloth used to make scenery

Set: The finished setup of scenery

Set dressing: Adding details such as props to a built set

Sheave: A pulley

Shutter: Plates inside a light used for beam shaping

Side light: Lighting from the side for dramatic effect

Sill iron: Metal bracing strip across bottom of a doorway in scenery

Sight line: Imaginary line from a viewer. Helps to establish what audience (and crew) are able to see

Special: Light used for special purpose. Sometimes used for spotlights that will pick out artists

Spike: To mark the stage, usually with tape, where objects are to be placed

Splitter: A cable with a plug and several sockets to split the input to several outputs. May be electrical or signal type

Spot block: Pulley fixed to the grid for a spot line

Spot line: A line added to the grid for a special purpose

Stage cloth: Stage floor covering

Stage left: The actor's left when facing the audience

Stage right: The actor's right when facing the audience

Stile: Side part of a flat

Strike: To remove objects from the stage

Swatch: Book containing samples of material or filters available

Tabs: Stage curtains situated in the wings

Teaser: Horizontal curtain used to mask lighting bars

Technical rehearsal: A run through to establish the technical aspects of a show

Throw line: A line used for tying flats or scenery together. Wraps around a throw line cleat

Thrust stage: Stage that projects out into the audience

Thunder sheet: Metal sheet that is shaken to sound like thunder

Toggle rail: Horizontal center rail in a flat

Tormentor: Flats used to mask offstage edges

Traps: Openings in stage floor with a trapdoor on the top. There is usually a trap room below for access to the traps

Tumbler: Bar fixed to the bottom of a cloth which is used to roll the cloth around. The cloth is rolled upwards using lines

Upstage: The area of the stage furthest from the audience

Variable lens: Light with a pair of lenses that can be used to adjust its focus

Wash: A general overall coverage of light often colored

Wash light: A light used to cover a general area. Usually uses a color filter

Winch: Device used for moving curtains, scenery or props. May be manual or motorized

Wing: Large flats used to mask side of stage

Wings: The areas at each side of the stage out of sight of the audience

Work lights: Sometimes called 'workers'. These are general lights used when working on stage when there is no performance taking place – such as when setting up or breaking down a show

Index